ISOTOPENPRAXIS

22. JAHRGANG

1986

AKADEMIE-VERLAG BERLIN

Isotopenpraxis

An international journal devoted to the application of isotopes and radiation
Founded by *J. Mühlenpfordt*, and *C. F. Weiss*
Edited at the Central Institute of Isotope and Radiation Research of the Academy of Sciences of the GDR

The journal deals with all aspects of the application of isotopes and radiation in the field of research, industry, agriculture and medicine. It is expected, that the papers give interesting and novel contributions to the application of isotopes and radiation, including isotope separation and production, radioisotope generators, syntheses of labelled compounds and radiopharmaceuticals nuclear instrument modules and electronic equipments, radiation detectors, radiometric gauges, safety engineering and radiation protection, radiation sources and radiation chemistry, cost benefit analysis, and questions of advanced training and postgraduate education.
In "Isotopenpraxis" orginal papers, short communications, reviews of important subject areas, book reviews, conference reports and news of interest to workers in this field are published. Papers may be written in English, German or Russian. It is assumed, that the papers have not been published previously.
Authors are held fully responsible for all information included in their papers.
Papers to be submitted for publication or books to be reviewed should be sent to

Professor Dr. sc. nat. H. Hübner
Editor-in-Chief
Karl-Marx-Platz
Hauptpostamt-P.O.Box 370
DDR-7010 Leipzig

Instructions for manuscript preparation are given on the third cover page.

Terms of subscription

Orders can be sent

- in the GDR: to Postzeitungsvertrieb, or to Akademie-Verlag, Leipziger Str. 3–4, DDR-1086 Berlin;
- in the other socialist countries: to a book-shop for foreign language literature or to the competent news-distributing agency;
- in the FRG and Berlin (West): to a book-shop or to the wholesale distributing agency „Kunst und Wissen“, Erich Bieber OHG, Wilhelmstr. 4–6, D-7000 Stuttgart 1;
- in the other Western European Countries: to Kunst und Wissen, Erich Bieber GmbH, Dufourstr. 51, CH-8008 Zürich;
- in other countries; to the international book- and journalselling trade; to Buchexport, Volkseigener Außenhandelsbetrieb der DDR, P.O. Box 160, DDR-7010 Leipzig; or to Akademie-Verlag, Leipziger Str. 3–4, DDR-1086 Berlin.

Herausgeber: Zentralinstitut für Isotopen- und Strahlenforschung der Akademie der Wissenschaften der DDR, DDR-7050 Leipzig, Permoserstr. 15
Verlag: Akademie-Verlag Berlin, Leipziger Straße 3–4, DDR-1086 Berlin, Fernruf: 2236221 oder 2236229; Telex-Nr.: 114420:
Bank: Staatsbank der DDR, Berlin, Kto.-Nr.: 6836-26-20712.
Chefredakteur: Prof. Dr. H. Hübner, Leipzig.
Anschrift der Redaktion der Zeitschrift Isotopenpraxis: Prof. Dr. Hübner, Karl-Marx-Platz, Hauptpostamt, PSF 370, DDR-7010 Leipzig; Fernruf: 28 10 18:
Veröffentlicht unter der Lizenznummer 1527 des Presseamtes beim Vorsitzenden des Ministerrates der Deutschen Demokratischen Republik.
Gesamtherstellung: VEB Druckerei „Thomas Müntzer“, DDR-5820 Bad Langensalza.
Erscheinungsweise: Die Zeitschrift „Isotopenpraxis“ erscheint monatlich. Die 12 Hefte eines Jahrganges bilden einen Band. Bezugspreis je Band 240,— DM zuzüglich Versandspesen: Preis je Heft 20,— DM. Der gültige Jahresbezugspreis für die DDR ist der Postzeitungsliste zu entnehmen.
Bestellnummer dieses Heftes: 1079/22

AN (EDV) 12831

Autoren- und Inhaltsverzeichnis

▶ Language of publication Publikationssprache

Sachverzeichnis

Band 22 · Januar 1986 **1**

isotopenpraxis

AKADEMIE-VERLAG · BERLIN

ISSN 0021-1915 Isotopenpraxis, Berlin 22 (1986) 1, 1—40

An unsere Autoren

Die Zeitschrift veröffentlicht Beiträge zu folgenden Themen: Kontrolle und Analyse radioaktiver und stabiler Isotope, Anwendung radioaktiver und stabiler Isotope in allen Gebieten der Forschung und Technik, der Industrie, Landwirtschaft und Medizin, Herstellung radioaktiver und stabiler Isotopen sowie von markierten Verbindungen, kernelektronische Geräte für die Isotopenanwendung (Dickenmesser, Dichtemesser u. ä.) Detektoren für isotopentechnische Arbeiten und den Strahlenschutz in Isotopenlaboratorien, Meßtechnik und Strahlenschutz, Dosimetrie in Isotopenlaboratorien, kerntechnische Geräte für die Isotopenanwendung (Beton, Bestrahlungsanlagen u. ä.), Strahlenchemie, Radiochemie, ökonomische und perspektivische Fragen, Fragen der Weiterbildung.

1. Aufnahme von Manuskripten

Zur Veröffentlichung angenommen werden nur druckreife, noch nicht anderweitig publizierte Beiträge in deutscher, englischer oder russischer Sprache zum Themenbereich der Zeitschrift und zwar

- Originalarbeiten
- Kurze Mitteilungen
- Aus der Praxis
- Zusammenfassende Berichte
- Tagungsberichte
- Informationen.

Sind der Beitrag oder Teile daraus bereits veröffentlicht oder ist eine anderweitige Veröffentlichung vorgesehen, so muß dies der Redaktion mit genauer Angabe der Zweitveröffentlichung mitgeteilt werden. Mit der Annahme des Manuskriptes behält sich der Verlag das Recht der Vervielfältigung, Verbreitung und Übersetzung innerhalb der gesetzlichen Schutzfrist vor. Auszugsweiser Nachdruck oder photomechanische u. a. Wiedergabe sind an eine schriftliche Genehmigung des Verlages gebunden. Manuskripte, Zeichnungen und Photos von veröffentlichten Arbeiten werden nur auf den besonderen und bis zum Zeitpunkt der Veröffentlichung mitgeteilten Wunsch des Autors zurückgesandt.

Bei der unverlangten Einreichung von Manuskripten wird empfohlen, eine Befürwortung eines der Herausgeber oder Redaktionsausschußmitglieder beizufügen.

Manuskripte sind einzureichen bei der Redaktion, 1086 Berlin Leipziger Str. 3—4

2. Form der Manuskripte

Die Autoren werden gebeten:

2 Manuskripte, eines davon in Original-Maschinenschrift, einseitig und mit 2zeiligem Abstand einzureichen; gut leserliche Photokopien sind zulässig; auf dem 1. Blatt (unter Weglassung der Bezeichnung)

- den Titel der Arbeit
- Vorname-Initialen und Namen sämtlicher Autoren
- gegebenenfalls die Institution, in der sie tätig sind
- eine kurze Zusammenfassung der Arbeit von höchstens 15 Zeilen
- Keywords in alphabetischer Reihenfolge und
- als Fußnote die Anschrift der Institution anzugeben;

wenn möglich, Zusammenfassung, Titel und Autoren *englisch* (bei deutschem oder russischen Manuskript) bzw. *deutsch* (bei englischem Manuskript) 2fach beizufügen;

Literaturhinweise (fortlaufend numeriert), Fußnoten, Tabellen, Abbildungen (in 2—3facher Veröffentlichungsgröße), Abbildungsunterschriften (getrennt von den Abbildungen und Zeichenerklärungen, Quellenangaben, falls erforderlich, und gegebenenfalls mit einem Vermerk für besondere Wünsche bezüglich der Abbildungsgröße), private und Dienstanschrift der Verfasser mit akademischen Titeln und ausgeschriebenen Vornamen auf besonderen Blättern ans Ende des Manuskriptes zu stellen;

Wünsche für den Kolumnentitel (Seitenüberschrift) bekanntzugeben;

Absätze durch Einzüge von mindestens 3 Anschlägen kenntlich zu machen; die Maßeinheiten des SI-Systems („Systeme International") zu verwenden; Korrekturen im Manuskript unmittelbar an der zu korrigierenden Stelle anzubringen;

Formeln deutlich lesbar und möglichst groß einzutragen;

Farbabbildungen und nachträgliche Änderungen oder Ergänzungen in Abbildungen zu vermeiden;

hervorzuhebende Textstellen zu unterstreichen, und zwar halbfetten Satz mit einer geraden schwarzen Linie, kursiven Satz (Schrägschrift) mit einer blauen Wellenlinie (alle Eigennamen). Sperrungen gestrichelt;

besondere Buchstaben durch farbige Unterstreichung zu kennzeichnen, und zwar Formelzeichen in

- Griechisch-Kursiv (α, β, ...) rot,
- Kursiv (*A*, *B*, ...) blau
- halbfetter Kursiv (**A**, **B**, ...) 2 Linien blau (Vektoren),
- halbfetter Grotesk (**A**, **B**, ...) 2 Linien violett (Tensoren);

bei Literaturangaben auf Vollständigkeit zu achten. In nachstehender Reihenfolge müssen angegeben sein:

a) bei *Zeitschriften*: Vorname-Initialen und Namen sämtlicher Autoren, Titel der Zeitschrift (nach „Periodica Chimica" oder möglichst wenig gekürzt), Bandnummer (unterstrichen), Jahreszahl (in Klammern), Heftnummer (nur bei Zeitschriften, die jedes Heft mit S. 1 beginnen) sowie die Seitenzahl, auf der die zitierte Arbeit beginnt.

b) bei *Büchern*: Initialen der Vornamen und Name des Autors (Herausgeber werden durch den Zusatz „ed." gekennzeichnet), voll ausgeschriebener Titel des Buches in der Originalsprache, evtl. Bandnummer, voll ausgeschriebener Name des Verlages, Nummer der Auflage (außer der ersten), Erscheinungsort(e), Erscheinungsjahr (hier nicht in Klammern setzen!) und evtl. Kapitelnummer, Seitenzahl u. ä.

c) bei *Sammelbänden, Tagungsberichten*: Es soll zusätzlich der Titel des zitierten Beitrags angegeben werden. Die Seitenzahl muß unbedingt angegeben werden. Aus dem Zitat muß hervorgehen, ob es sich um einen Tagungssammelband oder um einen noch nicht veröffentlichten Vortrag handelt. Zwischen Tagungsort und Erscheinungsjahr des entsprechenden Tagungsberichts ist zu unterscheiden.

3. Korrekturen

Der Autor erhält zwei Umbruchabzüge, von denen einer möglichst schnell korrigiert an die Redaktion zurückgereicht werden soll. Auf Wunsch des Autors kann die Korrektur auch von der Redaktion gelesen werden.

Durch nachträgliche Manuskriptänderungen verursachte Korrekturkosten gehen, sofern sie 10% der Satzkosten übersteigen, zu Lasten des Autors.

Der Autor erhält 50 Sonderdrucke kostenlos. Weitere Sonderdrucke können bis zur Rückgabe der Korrekturabzüge auf Kosten des Autors bei der Redaktion bestellt werden.

Bezugsmöglichkeiten

Bestellungen sind zu richten

- in der DDR an den Postzeitungsvertrieb unter Angabe der Kundennummer des Bestellenden oder an den AKADEMIE-VERLAG BERLIN, DDR-1086 Berlin, Leipziger Straße 3—4, PF-Nr. 1233;
- im sozialistischen Ausland an eine Buchhandlung für fremdsprachige Literatur oder an den zuständigen Postzeitungsvertrieb;
- in der BRD und Berlin (West) an eine Buchhandlung oder an die Auslieferungsstelle KUNST UND WISSEN, Erich Bieber OHG, Wilhelmstraße 4—6, D-7000 Stuttgart 1;
- in den übrigen westeuropäischen Ländern an eine Buchhandlung oder an die Auslieferungsstelle KUNST UND WISSEN, Erich Bieber GmbH. Dufourstraße 51, CH-8008 Zürich;
- im übrigen Ausland an den Internationalen Buch- und Zeitschriftenhandel; den Buchexport Volkseigener Außenhandelsbetrieb der Deutschen Demokratischen Republik, DDR-7010 Leipzig, Postfach 160; oder an den AKADEMIE-VERLAG BERLIN, DDR-1086 Berlin. Leipziger Straße 3—4.

Zeitschrift „Isotopenpraxis"

Herausgeber: Zentralinstitut für Isotopen- und Strahlenforschung der Akademie der Wissenschaften der DDR.
Verlag: Akademie-Verlag Berlin, DDR-1086 Berlin, Leipziger Straße 3—4, PF-Nr. 1233; Fernruf: 2236221 oder 2236229; Telex-Nr.: 114420;
Bank: Staatsbank der DDR, Berlin, Kto.-Nr.: 6836-26-20712.
Chefredakteur: Prof. Dr. H. Hübner, Leipzig.
Anschrift der Redaktion der Zeitschrift Isotopenpraxis: Prof. Dr. Hübner, DDR-7010 Leipzig, Karl-Marx-Platz, Hauptpostamt, PSF 370; Fernruf: 281018.
Veröffentlicht unter der Lizenznummer 1527 des Presseamtes beim Vorsitzenden des Ministerrates der Deutschen Demokratischen Republik.
Gesamtherstellung: VEB Druckerei „Thomas Müntzer", DDR-5820 Bad Langensalza.
Erscheinungsweise: Die Zeitschrift „Isotopenpraxis" erscheint monatlich. Die 12 Hefte eines Jahrganges bilden einen Band. Bezugspreis je Band 240,— DM zuzüglich Versandspesen; Preis je Heft 20,— DM. Der gültige Jahresbezugspreis für die DDR ist der Postzeitungsliste zu entnehmen.
Bestellnummer dieses Heftes: 1079/22/1.

AN (EDV) 12831
00900

Isotopenpraxis 22 (1986) 1, S. 1 bis 6

A Mathematical Approach to the Optimum Design of Gamma Irradiation Facilities I. Formulation of the Optimization Problem

E. Winkler

(Akademie der Wissenschaften der DDR, Zentralinstitut für Isotopen- und Strahlenforschung)[1])

Given a specified design the efficiency of a gamma irradiation facility can be improved by optimum adjustment of certain parameters (e.g. source positions, activity distribution, distance between the sources and the irradiation goods, dimensions of the irradiation container, etc.). The optimization problem is given by an objective function (efficiency, or throughput), which is to be maximized, and by constraints limiting the admissible range of the dose applied to the product. The general mathematical formulation of the problem is specified for different types of irradiation facilities (one or several conveyors, one or several containers in a stationary or an intermittent regime).

Aufbauend auf einer vorgegebenen Konstruktion kann der Wirkungsgrad einer Gammabestrahlungsanlage verbessert werden durch optimale Anpassung bestimmter Parameter (z. B. Quellenpositionen, Aktivitätsverteilung, Entfernung zwischen den Quellen und dem Bestrahlungsgut, Dimensionen des Bestrahlungsgutes usw.). Das Optimierungsproblem ist gegeben durch eine Zielfunktion (Wirkungsgrad oder Durchsatz), die maximiert werden soll, und durch Nebenbedingungen, die den zulässigen Bereich der dem Bestrahlungsgut zu applizierenden Dosis beschränken. Die allgemeine mathematische Formulierung des Problems wird für verschiedene Typen von Bestrahlungsanlagen konkretisiert (ein oder mehrere Förderbänder, ein oder mehrere Container in stationärer oder intermittierender Betriebsweise).

Keywords

computer calculations; gamma radiation; irradiation plants; optimization; radiation doses

Introduction

For more than 25 years methods for the processing of food and feed by means of gamma irradiation have been tested and applied in many countries. As a result of extensive investigations in 1980 an expert group from FAO/IAEA/WHO declared the treatment of food to be safe for health up to a dose of 10 Gy [1]. This recommendation renders possible a still broader application of irradiation methods which on the other hand requires a more intensive investigation of technological and economic problems.

In the GDR a pilot plant for the radiation treatment of onions (GBZ 81) was constructed in 1981 [2]. This facility uses the nuclide ^{60}Co as a source of gamma radiation. Its technological basis is rather simple. The onions (or, more generally, the irradiation goods) are continuously transported across the radiation field on the circular ring of a turning table. Given this design the distribution of the radiation dose in the irradiation goods depends on several parameters the values of which can be fixed within a certain range: the breadth of the circular ring, the thickness of the layer of the irradiation goods, the position of the sources, and the distribution of the ^{60}Co activity.

The experience gained from the design and the successful operation of this facility led to consideration about mathematical approaches to the optimum design of gamma irradiation plants. Here optimum design means the maximization of a characteristic for the productivity of the plant taking into consideration constraints regarding the energy dose to be applied to the irradiation goods. This paper described a method to solve this problem. In part I the optimization problem is formulated in terms of irradiation technology and of mathematics. Part II deals with the selection of an optimization algorithm and its implementation on the computer ES 1040 and gives examples of its application. The optimization method is considered mainly from the viewpoint of its practical applicability, while a more rigorous mathematical treatment has partly been omitted.

Until now the usefulness of the procedure has been proved for facilities working in a stationary regime (which is not economical for technological reasons [3]) as well as for continuously or intermittently working plants. The irradiation of fluids or other well mixed goods has not yet been taken into consideration. Facilities for the irradiation of food or feed are the starting point and the background for the procedure outlined here, but by no means do they limit its range of applicability.

The optimization problem

An optimization problem is specified by a mathematical model of the process under investigation, by constraints, and by an objective function to be minimized or maximized. In the context of irradiation facilities the mathematical model calculates the dose absorbed in different, sufficiently small volume elements of the irradiated substance taking into consideration the geometry of the problem and appropriate material constants. Constraints limit the energy dose to be applied to the irradiation goods or the dose rate as well as admissible ranges for the design parameters, or they fix certain relations between the parameters. An appropriate objective function is given by a quantity which characterizes the capacity of the irradiation plant and which can form a basis for economic calculations.

[1]) Address: DDR-7050 Leipzig, Permoser Str. 15

The capacity of a plant can be improved in two steps:

Fundamental changes in its design may be very efficient but they cannot be formalized in a mathematical algorithm. Above all they are based on technical knowledge and experience. Among such changes there are also measures taken to improve the time utilization coefficient f_t which gives the ratio of the actually utilized to the total time of irradiation.

If a specified design is already given a number of continuous or discrete parameters remain open. Generally there are admissible ranges for these parameters depending on the design.

This paper deals with a section of the second optimization step, the appropriate adjusting of continuously variable parameters by means of a mathematical procedure. Discrete parameters are only indirectly taken into consideration. In general, there exist only a small number of possible values for these quantities. It is possible, therefore, to fix a discrete parameter value (or a set of values) each time and to calculate the continuous ones under this supposition.

A quantity which is of direct interest for economic calculations is the throughput, L given by

$$L = \Delta m / \Delta t \qquad (1)$$

where Δm means the mass of the substance which can be irradiated in the time interval Δt if $t_t = 1$. This quantity is used here as the objective function for the optimization.

The throughput L is calculated on the basis of a given source activity of the plant. There exists a linear relationship between L and the quality factor Γ which is of great importance in irradiation technology [4, 5]. The latter quantity is defined by the ratio of the energy usefully absorbed by the irradiation goods in a given time intervall to the total energy emitted by the sources in the same time interval

$$\Gamma = \frac{L \cdot D_{\min}}{K \cdot Q} \qquad (2)$$

where

L throughput, kg/h

$D_{\min}$ minimum dose, Gy, required to attain the purpose of the irradiation

Q activity of the ^{60}Co sources (Bq or Ci)

K a conversion constant ($K = 1{\cdot}44 \cdot 10^{-12}$ kGy · kg/h · Bq, or 5·34 Mrd · kg/h · kCi for ^{60}Co)

The quantity Γ defined by (2) is also called efficiency factor. But this term is confusing, for it is also used for another quantity, η, which gives the ratio of the total energy absorbed by the irradiation goods in a given time interval to the total energy emitted by the sources [6]

$$\eta = \frac{L \cdot \bar{D}}{K \cdot Q} \qquad (3)$$

$\bar{D}$ is the statistical mean of the dose absorbed by the volume elements of the irradiated substance (the overall average dose). The quantity η is not well suited for optimization. It depends not only on the throughput L but also on a change in the dose distribution within the admissible range and hence in the mean $\bar{D}$, possible even at the expense of the throughput.

The maximization of the objective function L is restricted by conditions as to the admissible range of dose. Most unambiguous is the establishment of definite limits $D_{\min}$ and $D_{\max}$. In other words, two constraints are given:

– A given minimum dose $D_{\min}$ must be attained over the total volume of the irradiated substance.
– The dose absorbed in any volume element must be greater than $D_{\min}$ only by a given factor $U_{\max}$, the permissible maximum overdose.

These constraints are combined by

$$1 \leqq U = D/D_{\min} \leqq D_{\max}/D_{\min} = U_{\max} \qquad (4)$$

Another formulation of the constraints as to the dose is somewhat less stringent: given limits $D_{\min}$ and $D_{\max}$ must be observed for a mass-percentage of the irradiated substance of at least 97·5%. Taking these possible violations of the dose limits into consideration will lead to very time-consuming calculations. In each step of the optimization algorithm a detailed determination of the dose distribution would be necessary, while if definite limits are given the dose need only be calculated for a small number of field points, at the best for two. It turns out to be better to observe the constraints of eq. (4) during the calculation in order to have a reserve for the inevitable discrepancies between theoretical considerations and experimental results.

Observing a given value for $\bar{D}$ leads to the same difficulties as above and is not considered in the following formulations. Also additional constraints as to the dose rate are neglected though they may result, e.g., when seeds are to be irradiated.

In summary, the optimization problem reads as follows:

– The throughput L (e.g. kg/h) is to be maximized (objective of the optimization).
– The dose applied to volume elements of the irradiation goods is limited by minimum and maximum values $D_{\min}$ and $D_{\max}$, Gy (constraints).

An irradiation plant can only be optimized for a given substance to be irradiated, characterized by its density. Yet not only the throughput depends on the density (detailed investigations in [3])but also the combination of parameter values giving the maximum throughput. This is to be taken into account when an irradiation facility is planned.

Mathematical formulation of the optimization problem

The optimization problem is mathematically given by

$$L = L(p_1, \ldots, p_n) \rightarrow \text{Max} . \qquad (5\text{a})$$

$$D_{\min} \leqq D(x, y, z, p_1, \ldots, p_n) \leqq D_{\max} \qquad (5\text{b})$$

$$\text{for} \quad x, y, z \ni V$$

$$\sum_j a_{jl} - b_j = 0 \qquad (5\text{c})$$

$$\text{with} \quad l = 1, \ldots, n \qquad j = 1, \ldots, m \qquad m < n$$

$$c_{1l} \leqq p_l \leqq c_{2l} \qquad (5\text{d})$$

Eq. (5a) gives the objective function, (5b) the dose constraints; additionally it may be necessary to take into consideration linear relationships between the parameters p_l (5c) or limits for the parameters (5d). The throughput does not necessarily depend directly on all parameters, and some parameters only limit the irradiation volume V without appearing explicitely in the dose constraints.

As formulated in eq. (5b) the dose constraints are difficult to use because of the continuous coordinates x, y, z (which stand for any 3 coordinates describing the position of a point in space). It can be shown [7] that instead of the continuous model (5b) also a simpler one, got by the introduction of discrete field points, can be investigated:

$$D_{\min} \leqq D(x_i, y_i, z_i, p_l) \leqq D_{\max} \qquad (6)$$
$$x_i, y_i, z_i \ni V$$
$$i = 1, \dots n_k$$

It is always possible to select a set of field points as representative as necessary.

The calculation of the applied dose is based on the determination of the dose rate $\dot{D}_i$ at a given field point (x_i, y_i, z_i), given by

$$\dot{D}_i = \frac{f \cdot F \cdot Q}{(x_i - x_Q)^2 + (y_i - y_Q)^2 + (z_i - z_Q)^2} \times e^{-\left(\frac{\mu}{\varrho}\right)\varrho d_i'} \cdot B\left(\frac{\mu}{\varrho}\,\varrho d_i'\right) \quad [\mathrm{Gy/h}] \qquad (7)$$

with

$$B\left(\frac{\mu}{\varrho}\,\varrho d_i'\right) = A e^{-\alpha_1 \frac{\mu}{\varrho}\varrho d_i'} + (1 - A)\, e^{-\alpha_2 \frac{\mu}{\varrho}\varrho d_i'}$$

and

- F — dose constant ($F = 12930$ r · cm²/h · Ci or $F = 3.495 \cdot 10^{-7}$ r · cm²/h · Bq for ^{60}Co)
- f — correction constant for converting the theoretical activity Q into the effective one and from roentgen (r) into gray (Gy). For the calculations in part II $f = 0.0067$ was taken.
- x_i, y_i, z_i — coordinates of the field point i
- x_Q, y_Q, z_Q — coordinates of the source with activity Q
- d_i' — effective thickness of the material layer penetrated by the radiation between source and field point i
- μ — linear mass absorption coefficient
- ϱ — density of the irradiated material
- B — build-up factor with the material constants A, α_1, α_2. For $\tau = \mu d_i' \approx 2$ a linear approximation of B is sufficient

The general formalism (5) ... (7) was used for the calculation of different types of facilities. In each case the equations take on a specified form.

Conveyor

The irradiation goods continuously pass the sources with the velocity v on a conveyor (the radiation absorption of which does not directly appear in the formulas below but is included in the coefficient f). The sources are at some position above and below the conveyor unit. Fig. 1 shows a cross section with 2 point sources. For a mathematical description two coordinate systems are introduced. The first on (y^*, z^*) moving with the irradiation goods is defined only in the y-z-plane as because of the integration over the dose rate along the x-direction the applied dose does not depend on x. The origin is marked by (•) in Fig. 1. The second one fixed in space (x, y, z) is appropriately defined in such a way that

$$\left.\begin{aligned} y &= y^* \\ z &= z^* \end{aligned}\right\} \qquad (8)$$

The coordinate value $x = 0$ should be defined by the position of the sources. These are denoted by x_{Qj}, y_{Qj}, z_{Qj} in the fixed system. n_1 sources are above, n_2 below the conveyor. For a conveyor facility the general equations now read

$$L = 2sdv\varrho \to \mathrm{Max}\,. \qquad (9\mathrm{a})$$

$$D_{\min} \leqq D_i(y_i^*, z_i^*, Q_j, x_{Qj}, y_{Qj}, z_{Qj}, d, s, h, v) \leqq D_{\max}$$
$$-s \leqq y_i^* \leqq +s \qquad 0 \leqq z_i^* \leqq d \qquad (9\mathrm{b})$$

with

$$D_i(y_i^*, z_i^*) = D_i(y_i, z_i) = \frac{1}{v}\int_{x_{\min}}^{x_{\max}} \dot{D}_i(x, y_i, z_i)\, dx$$

$$\dot{D}_i = \sum_{j=1}^{n_1+n_2} \dot{D}_{ij}$$

$$\dot{D}_{ij} = \frac{fFQ_j}{d_{ij}^2}\, e^{-\frac{\mu}{\varrho}\varrho d_{ij}'} \cdot B\left(\frac{\mu}{\varrho}\,\varrho d_{ij}'\right)$$

$$d_{ij}^2 = (x_i - x_{Qj})^2 + (y_i - y_{Qj})^2 + (z_i - z_{Qj})^2$$

$$d_{ij}' = d_{ij}\,\frac{|z_{Qj} - z_i| - h}{|z_{Qj} - z_i|}$$

$$z_{Qj} = h + d \qquad (9\mathrm{c})$$

for the first group of sources

$$|z_{Qj}| = h$$

for the second group of sources

$$y_{Q_1} = -y_{Q_2}, \quad \text{etc.}$$
$$\sum_j Q_j = Q_{\text{tot.}}$$
$$|y_{Qj}| \leqq s \qquad (9\mathrm{d})$$

The parameter s is best treated as a discrete quantity.

Several conveyors

The utilization of the radiation will be increased if the irradiation goods pass the sources several times at different distances. An example of such a facility consisting of 3 conveyors is shown in Fig. 2 in cross and longitudinal section. The irradiation goods pass the sources on the highest conveyor, fall on the next one, and the irradiation is repeated. In principle the irradiation on each conveyor is also described by eqs. (9a) ... (9d). Additionally, absorption of radiation by the central layer must be taken into account

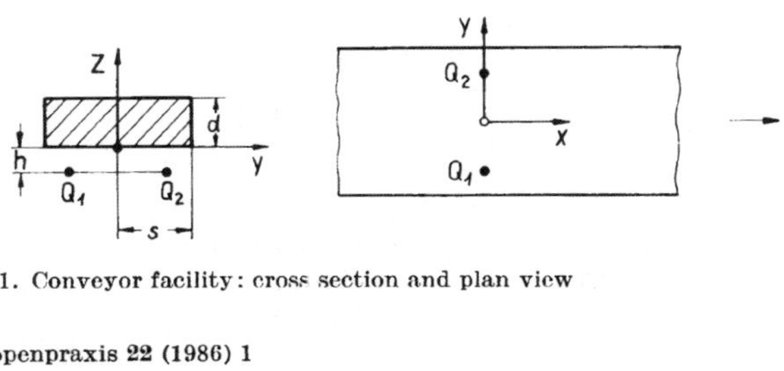

Fig. 1. Conveyor facility: cross section and plan view

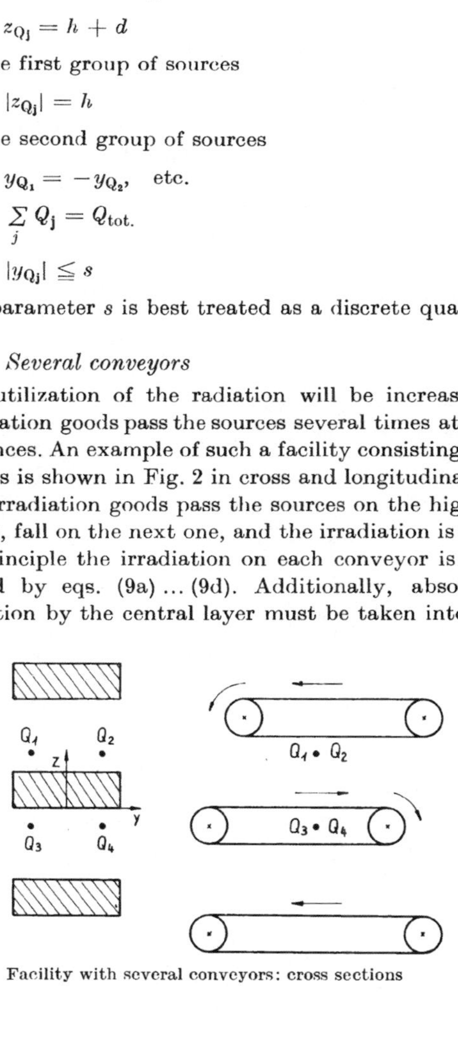

Fig. 2. Facility with several conveyors: cross sections

when the dose applied to the material on the top and bottom conveyors is calculated. — When the total system is considered it can be assumed that the irradiated material will be well mixed when falling from one conveyor on the next one. As a result of this mixing the calculation of the resulting dose distribution is considerably more difficult than for a single conveyor. Examples show that the dose distribution which is rather asymmetric for most types of facilities approaches a normal distribution as the number of conveyors increases. Hence dose values near D_{min} and D_{max} occur less frequently. But it cannot strictly be excluded that a certain volume element of the irradiated substance obtains either the proportional minimum dose D_{min_j} or maximum dose D_{max_j} on each conveyor j, and therefore the values

$$\left.\begin{aligned} D_{max} &= \sum_{j=1}^{n_C} D_{max_j} \\ D_{min} &= \sum_{j=1}^{n_C} D_{min_j} \end{aligned}\right\} \qquad (10)$$

limit also the resulting distribution. — The individual D_{min_j} and D_{max_j} differ from each other dependent on the distance of each conveyor to the sources. It can be assumed that for an optimum facility also the individual maximum overdoses $U_{max_j} = D_{max_j}/D_{min_j}$ are different. But in order to optimize a facility under this general assumption a computer program for the laborious calculation of the dose distribution of the complex arrangement must be written. Under a simplified (probably nonoptimum) assumption. the maximum permissible overdose U_{max} of the resulting distribution must be observed for each conveyor:

$$U_{max_j} = U_{max} \qquad (j = 1, \ldots, n_C) \qquad (11)$$

The main part of the total dose (ca. 70%) is applied to the product on the central conveyor. Therefore the multiconveyor facility can be optimized step by step. At first only the central conveyor is considered: the optimum thickness of the layer, positions of the sources, and the distribution of the activity are calculated (with the conveyor width as a discrete parameter). Now the velocities of the outer conveyors relative to that of the central one are fixed (the determination of the optimum value for these ratios would again lead to the simultaneous treatment of the total facility). Given these relative velocities and the requirement of equal throughput on each conveyor the thickness of the layers on the outer conveyors can easily be calculated. At last only the optimum distances of these conveyors are to be determined, just as the absolute velocities of all conveyors.

Turning table

The material to be irradiated can pass the sources also on a turning table with an angular velocity ω, 1/h. This possibility was used in the irradiation plant GBZ 81 at Weideroda [2]. Fig. 3 gives a plan view and a cross section of the plant. As in the case of the conveyor unit 2 coordinate systems are defined for the calculation. The moving system (coordinates y^*, z^*) is defined as above (see also Fig. 3a), while for the fixed one cylindrical coordinates (r, φ, z) are introduced the origin of which is placed at the centre of the table on its surface. The transformation is given by

$$\left.\begin{aligned} y^* &= r - R \\ z^* &= z \end{aligned}\right\} \qquad (11)$$

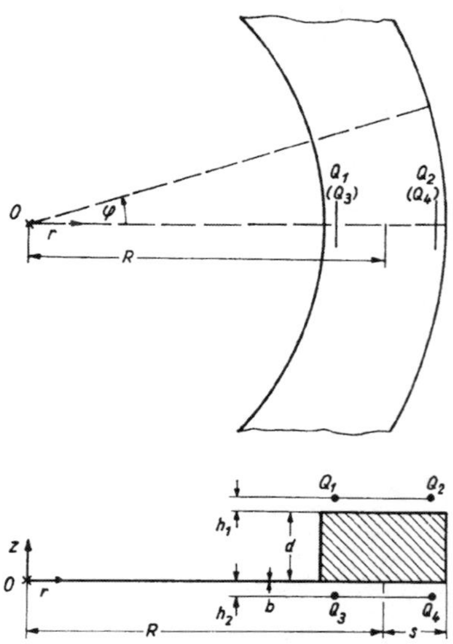

Fig. 3. Turning table facility: cross section and plan view

In contrast to the conveyor belt the radiation absorption in the table (iron of thickness b) must be taken into account.

The optimization problem for the turning table is described by

$$L = 2sd\omega\varrho R \to \text{Max.} \qquad (13a)$$

$$\begin{aligned} &D_{min} \leqq D_i(y_i^*, z_i^*, Q_j, r_{Q_j}, \varphi_{Q_j}, z_{Q_j}, h_1, h_2, \omega) \leqq D_{max} \\ &-s \leqq y_i^* \leqq s \qquad\qquad (13b) \\ &0 \leqq z_i^* \leqq d \end{aligned}$$

with

$$D_i(y_i^*, z_i^*) = D_i(r_i, z_i) = \frac{1}{\omega} \int_{\varphi_{min}}^{\varphi_{max}} \dot{D}_i(r_i, z_i, \varphi)\, d\varphi$$

$$\dot{D}_i(r_i, \varphi_i, z_i) = \sum_{j=1}^{n_1} \dot{D}_{ij}^{(1)} + \sum_{j=1}^{n_2} \dot{D}_{ij}^{(2)}$$

$$\dot{D}_{ij}^{(1)} = \frac{fFQ_j}{d_{ij}^2} e^{-\frac{\mu}{\rho_w} d'_{ij}\rho} \left\{ A' e^{-\alpha'_1 \frac{\mu'}{\rho_w} d'_{ij}\rho} + (1 - A') e^{-\alpha'_2 \frac{\mu'}{\rho_w} d'_{ij}\rho} \right\}$$

first group of sources (above the table)

$$\dot{D}_{ij}^{(2)} = \frac{fFQ_j}{d_{ij}^2} \times e^{-\left(\frac{\mu}{\rho_w} d'_{ij}\rho + \mu'' d''_{ij}\right)} \left\{ A' e^{-\alpha'_1 \frac{\mu'}{\rho_w} d'_{ij}\rho} + (1 - A') e^{-\alpha'_2 \frac{\mu'}{\rho_w} d'_{ij}\rho} \right\} \cdot \left\{ A'' e^{-\alpha''_1 \mu'' d''_{ij}} + (1 - A'') e^{-\alpha''_2 \mu'' d''_{ij}} \right\}$$

second group of sources (below the table)

$$d_{ij}^2 = r_i^2 + r_{Q_j} - 2 r_i r_{Q_j}^2 \cos(\varphi_{Q_j} - \varphi_i) + (z_{Q_j} - z_i)^2$$

$$d'_{ij} = d_{ij} \frac{|z_{Q_j} - z_i| - h}{|z_{Q_j} - z_i|} \qquad h = \begin{cases} h_1 \\ h_2 \end{cases}$$

$$d''_{ij} = b \frac{(d_{ij} - d'_{ij})}{|z_{Q_j}|}$$

(μ', A', α'_1, α'_2) material constants for water
(μ'', A'', α''_1, α''_2) material constants for iron

ϱ density of the irradiation goods
ϱ_w density of water

$$
\begin{aligned}
& z_{Qj} = h_1 + d \quad \text{(first group of sources)} \\
& z_{Qj} = -h_2 \quad \text{(second group of sources)} \\
& \sum_{j=1}^{n} Q_j = Q_{tot} \\
& R - s \leqq r_{Qj} \leqq R + s \qquad (13\text{d}) \\
& \varphi_{Qj} = 0 \quad \text{(or another fixed position)}
\end{aligned}
$$

The parameters s and R may again be treated as discrete variables.

The dose field in the (y^*, z^*) — plane is irregular. It is not symmetric about the axis $y^* = 0$ because of the different radial velocities of the table at different coordinates r, neither is it symmetric about the axis $z^* = d/2$ because of the additional absorption by the iron table. This makes the calculation of a representative dose field rather cumbersome.

For all cases treated up to now the time utilization coefficient, as a consequence of the continuous irradiation regime, is nearly unity, and it is only diminished by breakdowns.

Container; stationary regime

Many irradiation goods must be packed into containers. As a simple possibility Fig. 4 shows such a unit with one source and one container which is irradiated for the time T. The two coordinate systems have their origin in the source (x, y, z) and at the centre of the container (x^*, y^*, z^*), respectively. The transformation is given by

$$
\left.
\begin{aligned}
x &= x^* + A_x + h \\
y &= y^* \\
z &= z^*
\end{aligned}
\right\} \qquad (14)
$$

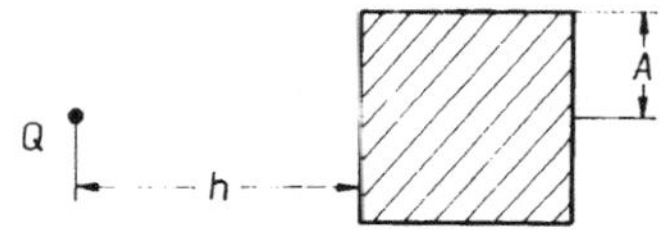

Fig. 4. Container facility: cross section

The throughput of the unit is optimized according to

$$L = 8 A_x A_y A_z \varrho / T \qquad (15\text{a})$$

$$D_{min} \leqq D_i(x_i^*, y_i^*, z_i^*, h, T) \leqq D_{max} \qquad (15\text{b})$$

$$-A_x \leqq x_i^* \leqq A_x$$
$$-A_y \leqq y_i^* \leqq A_y$$
$$-A_z \leqq z_i^* \leqq A_z$$

with

$$D_i(x_i^*, y_i^*, z_i^*) = T \dot{D}_i(x_i, y_i, z_i)$$

$$\dot{D}_i = \frac{fFQ}{d_i^2} e^{-\frac{\mu}{\varrho_w} d'_{i\varrho}} \left\{ A e^{-\alpha_1 \frac{\mu}{\varrho_w} d'_{i\varrho}} + (1 - A) e^{-\alpha_2 \frac{\mu}{\varrho_w} d'_{i\varrho}} \right\}$$

$$d_i^2 = x_i^2 + y_i^2 + z_i^2$$

$$d'_i = d_i \frac{|x_i - h|}{|x_i|}$$

Additional containers may be placed around the central source, and they may be turned after a given time by a given angle (e.g. 180°). In this way the throughput is considerably increased. Nevertheless such a facility has a dead time during the transport of the containers into and out of the irradiation chamber. Due to the stationary regime the time utilization coefficient is considerably less than 1.

Container; intermittent regime

When a container facility is operated in an intermittent regime its time utilization coefficient can be approximated to that of a continuously working unit. Intermittent regime means that the irradiation containers pass the facility with interruptions of the movement. After a certain time (the unit time T) during which a container is irradiated on one position all containers are transported to their next position while retaining their orientation in space (see Fig. 5). At the end of the irradiation (i.e. after one cycle with the cycle time $T_c = n \cdot T$, where n is the number of positions) a container has turned all its side walls to the sources. The coefficient f_t is negatively influenced only if the time needed to change the position or to move into or out of the irradiation chamber is not small compared to the unit time T [8].

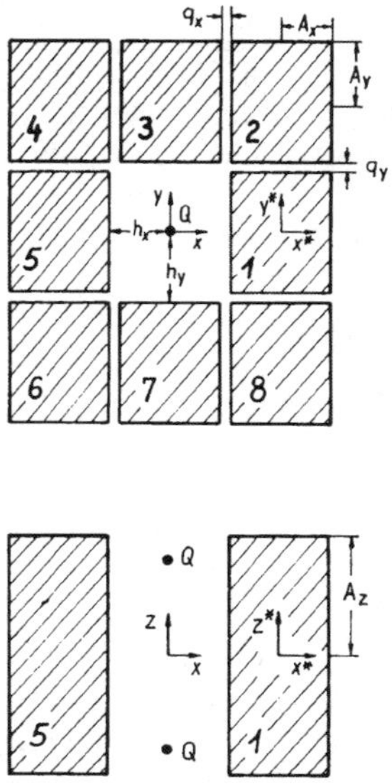

Fig. 5. Facility with several containers (intermittent regime): plan view and cross section

The two coordinate systems necessary for the calculation of the dose are transformed in the following way

$$
\left.
\begin{aligned}
x &= x^* + x_0 \\
y &= y^* + y_0 \\
z &= z
\end{aligned}
\right\} \qquad (16)
$$

The coordinates x_0 and y_0, the origin of the moving system in terms of the fixed system, are given in Tab. 1 for each position of the container.

Tab. 1. The coordinates for the centre of the container (see Fig. 5)

container position k	x_0k	y_0k
1	$h_x + A_x$	0
2	$2A_x + q_x$	$2A_y + q_y$
3	0	$h_y + A_y$
4	$-2A_x - q_x$	$2A_y + q_y$
5	$-h_x - A_x$	0
6	$-2A_x - q_x$	$-2A_y - q_y$
7	0	$-h_y - A_y$
8	$2A_x + q_x$	$-2A_y - q_y$

Before the optimization problem is formulated two problems concerning the determination of the distance across which the radiation penetrates the material must be mentioned in some detail. The first problem arises from the gaps (q_x and q_y) between the containers. To simplify programming it was assumed that they are also filled with material. The error in the determination of the dose rate remains insignificant if the gaps are narrow (e.g. 10% of $2 \cdot A_x$ or $2 \cdot A_y$). The second problem concerns the determination of the coordinates (x_s, y_s) of the correct point of intersection between the radius from the source j to the field point i and the planes $x = \pm h_x$ and $y = \pm h_y$ the intersections of which with the plane $z = 0$ are denoted by $g_1 \dots g_4$. The straight line G between the source and the field point in the plane $z = 0$ is given by

$$G \equiv -(y_i - y_{Qj})\, x + (x_i - x_{Qj})\, y + x(y_i - y_{Qj}) - y_{Qj}(x_i - x_{Qj}) = 0 \tag{17}$$

In the general case there exist points of intersection of G with all four straight lines $g_1 \dots g_4$. To select the correct one it is taken into account that at most two of them lie in the admissible range of G. But only one of these two also lies in the admissible range of one of the g_i. The case when G runs parallel to a pair of the g_i must be considered separately.

The optimization problem can now be written down as follows:

$$L = 8 A_x A_y A_z \varrho / T \rightarrow \text{Max.} \tag{18a}$$

$$D_{\min} \leqq D_i(x_i^*, y_i^*, z_i^*, x_{Qj}, y_{Qj}, z_{Qj}, h_x, h_y, q_x, q_y, T) \leqq D_{\max} \tag{18b}$$

$$-A_x \leqq x_i^* \leqq A_x$$

$$-A_y \leqq y_i^* \leqq A_y$$

$$-A_z \leqq z_i^* \leqq A_z$$

with

$$D_i(x_i^*, y_i^*, z_i^*) = T \cdot \sum_{j=1}^{n_Q} \sum_{k=1}^{8} \dot{D}_{ij}(x_i^* + x_{0k}, y_i^* + y_{0k}, z_i^*) = T \sum_{j=1}^{n_Q} \sum_{k=1}^{8} \dot{D}_{ij}(x_{ik}, y_{ik}, z_i)$$

$$\dot{D}_{ij}(x_{ik}, y_{ik}, z_i) = \frac{f F Q_j}{d_{ijk}^2}\, e^{-\frac{\mu}{\rho_w} d'_{ijk} \rho} \left\{ A e^{-\alpha_1 \frac{\mu}{\rho_w} d'_{ijk} \rho} + (1 - A)\, e^{-\alpha_2 \frac{\mu}{\rho_w} d'_{ijk} \rho} \right\}$$

$$d_{ijk}^2 = (x_{ik} - x_{Qj})^2 + (y_{ik} - y_{Qj})^2 + (z_{ik} - z_{Qj})^2$$

$$d'_{ijk} = d_{ijk} \frac{\sqrt{(x_{sk} - x_{ik})^2 + (y_{sk} - y_{ik})^2}}{\sqrt{(x_{Qj} - x_{ik})^2 + (y_{Qj} - y_{ik})^2}}$$

$$\sum_{j=1}^{n_Q} Q_j = Q_{tot} \tag{18c}$$

Further conditions as to the parameters arise if there are more than one sources and the sources are to be arranged symmetrically.

Conclusion

Optimization problems often arise in the medical application of the irradiation technique (e.g. *Langheim* [9], *Franke* [10]). The author owes these and other papers suggestions concerning the statement of the problem given here as well as its solution. The selection of an appropriate algorithm, its implementation on a computer and examples of its application are presented in part II of this paper [11].

Received April 3, 1985
Accepted in revised form May 9, 1985

References

[1] FAO/IAEA/WHO Expert Committee "Wholesomeness of Irradiated Food", World Health Organisation, Geneva 1981
[2] *R. Döllstädt, Ch. Grahn, G. Hübner, J. Reinhardt, H.-W. Schulze, H. Sprinz, E. Winkler*, ZfI-Mitt. *70* (1983) 178
[3] *S. Rulis*, Thesis, TU Dresden, 1975
[4] *W. Suckow*, Thesis, TU Dresden, 1968
[5] *K. H. Tetzlaff*, Atomkernenergie-Kerntechnik *34* (1979) 305
[6] *A. Brynjolfsson*, in: Factors Influencing the Economical Application of Food Irradiation. Wien: IAEA 1973 (STI/PUB/331), p. 13
[7] *U. Ebert*, Optimale Auslegung von Bestrahlungsplänen (Schriftenreihe der Universität Münster). Münster 1977
[8] *G. Hübner*, ZfI-Mitt. *98* (1984) 578
[9] *F. Langheim*. in: Medizinische Physik (*W. J. Lorenz*, ed.), Vol. 2. 1977, p. 549
[10] *D. S. Franke*, Radiobiol.-Radiother. *21* (1980) 68
[11] *E. Winkler*, Isotopenpraxis *22* (1986) 7

Isotopenpraxis 22 (1986) 1, S. 7 bis 11

A Mathematical Approach to the Optimum Design of Gamma Irradiation Facilities II. A Computer Program for Nonlinear Optimization and Its Application

E. Winkler

(Akademie der Wissenschaften der DDR, Zentralinstitut für Isotopen- und Strahlenforschung, Leipzig)[1])

To solve the problem of the optimization of gamma irradiation facilities given by an objective function (throughput), which is to be maximized, and by constraints (in terms of nonlinear inequalities and of linear equations) a computer program was developed based on the evolution method, a combination of a stochastic search and an aim-oriented step-by-step procedure. The applicability of the alogrithm implemented on an ES 1040 computer to the maximization of the throughput as well as to the improvement of the dose field homogeneity is demonstrated by different examples especially related to the irradiation of onions.

Um das Problem der Optimierung von Gammabestrahlungsanlagen zu lösen, welches durch eine zu maximierende Zielfunktion (Durchsatz) und durch Nebenbedingungen (nichtlineare Ungleichungen und lineare Gleichungen) gegeben ist, wurde ein Computerprogramm erarbeitet, welches auf dem Evolutionsverfahren beruht, einer Kombination einer stochastischen Suche und eines zielorientierten schrittweisen Verfahrens. Der Algorithmus wurde auf einem Rechner ES 1040 implementiert. Seine Anwendbarkeit sowohl bei der Maximierung des Durchsatzes als auch bei der Verbesserung der Dosisfeldhomogenität wird durch verschiedene Beispiele gezeigt, welche sich insbesondere auf die Bestrahlung von Zwiebeln beziehen.

Keywords

computer calculations; computer codes; gamma radiation; irradiation plants; optimization; radiation doses

Introduction

In part I of this paper [1] it was discussed in which way the capacity of a gamma irradiation facility can be enlarged on the assumption that the main features of the design are given. An optimization problem was formulated with the throughput L of the facility as an objective function to be maximized and with constraints as to the radiation dose D applied to the irradiation goods at different field points (x_i, y_i, z_i) and as to the design parameters p_l. Mathematically the problem reads as follows (cf. I, eqs. (5) and (6)):

$$L = L(p_1, \ldots, p_n) \rightarrow \text{Max.} \tag{1a}$$

$$D_{\min} \leqq D_i = D(x_i, y_i, z_i, p_l) \leqq D_{\max} \tag{1b}$$

$$(\text{or} \quad 1 \leqq U_i \leqq U_{\max} \quad \text{with} \quad U_i = D_i/D_{\min})$$

$$i = 1, \ldots, n_k$$

$$\sum_l a_{jl} p_l - b_j = 0 \tag{1c}$$

$$\text{with} \quad l = 1, \ldots, n \quad j = 1, \ldots, m \quad m < n$$

$$c_{1l} \leqq p_l \leqq c_{2l} \tag{1d}$$

The model functions which calculate the dose D depend on the special desgin of the facility. Examples are discussed in part I (eqs. (9), (13), (15), (18)).

Methods for the solution of the optimization problem

Eqs. (1a) ... (1d) represent an optimization problem with a nonlinear objective function as well as nonlinear or linear inequalities and equations as constraints. This optimization problem can be solved by a definite set of values for the parameters both in the objective function and in the constraints. To find this combination of values representing a point in the parameter space an appropriate algorithm must be used, i.e. an instruction that generates a sequence of points which approximates the solution [2]. For the selection of such an algorithm some peculiarities of the problem (1) have to be taken into account:

— The objective function depends directly only on a subset of the parameters; especially the positions of the sources, the distribution of the activity, and the position of the absorbing material do not appear in the different formulations of eq. (1a) (see I/(9a), (13a), (15a), (18a)).

— Also the constraints as to the dose (1b) depend on certain parameters (the dimensions of the irradiation goods) only in an indirect way.

— Some quantities appear in the constraints as nonlinear parameters. The connexion is especially complicated if it is formed by integrals the solution of which cannot be given in a closed form (I/(9b), (13b)).

— The dose constraints are partly equations, partly inequalities. While inequalities only reduce the size of the admissible range of the parameter space each equation lowers the number of independent parameters by one. But in general it is not known in advance for which coordinates (x_i, y_i, z_i) the sign of equality is valid. (If we have this knowledge, at least partially, the amount of calculations to characterized the dose field can be reduced considerably).

— The mathematical formulation of the optimization problem still contains a group of linear equations between the parameters (eq. (1c)). By their help it would be possible to substitute some of the parameters in the objective function or in the dose constraints depending on the special pro-

[1]) Address: DDR-7050 Leipzig, Permoserstr. 15

blem. But as a computer program as general as possible was aimed at these equations were left at their place.

— It is supposed to be a necessary condition that the objective function reaches its maximum with a combination of parameter values for which also

$$\max (U_i) = U_{max} \quad (2)$$

is valid where U_{max} is the maximum permissible overdose. To prove this assumption with mathematical rigour is not a trivial task. If the dimensions of the irradiation goods (e.g. the thickness of the layer on a conveyor, the dimensions of a container) belong to the free parameters of a problem then this assumption is immediately obvious as, starting from a parameter combination which results in $\max (U_i) < U_{max}$, it is possible to enlarge these dimensions in such a way that also the throughput increases without the dose constraints being violated. With other problems special discussions would be necessary; at least it can be stated that the calculations performed up to now do not contradict the assumption.

As for important cases the dose constraints are given by an integral to be solved numerically it was avoided from the first to use any of the optimization methods which make use of the differentiation of the objective function and the constraints by their parameters. Therefore in a final choice remain the so called direct methods which approximate the solution without making use of the analytical properties of the objective function and the constraints [2, 3]:

— When a stochastic search for the optimum is performed (*Monte-Carlo*-method, pure random search) a certain number of points in the parameter space are selected randomly, the objective function and the dose field with these parameters are calculated, and that combination of parameters is taken as the solution which gives the maximum of the objective function without violating the dose constraints. No information as to the precision of the location of the maximum is obtained; the amount of calculations increases exponentially with the number of parameters.

— The pattern search method is somewhat more favorable as to the computation time, but especially as to the planning of the calculations. The precision with which the optimum of the objective function is located depends strongly on the width of the meshes of the search pattern. In certain cases, if the number of parameters is small ($n \leq 4$) this method may be well suited also for microcomputers while for growing n the time of calculation increases in an unjustifiable manner (again exponentially). *Jank* and *Meister* [4] demonstrate this by a drastic example. The pattern search method was successfully applied by *Hargittai* and *Stenger* [5] for the planning of a poppy irradiator consisting of a vertical ring-shaped tube with the sources arranged around the tube and on its axis.

— The pattern search method is still more efficient if the field points are known where the dose reaches its maximum and minimum values. As already outlined above the m equations ($m \geq 2$) which form a part of the dose constraints (1 b) can actually be written down in this case and hence the number n of free parameters is reduced. In the pattern search it is necessary to fix combinations of only $(n - m)$ parameters and to determine the remaining m parameters by solving the (nonlinear) equations of the dose constraints. Finally the objective functions are compared in the above manner. Such a procedure was applied when the onion irradiator GBZ 81 at Weideroda was designed. Compared with a general pattern search method this special case has the advantage of eq. (2) being guaranteed from the first, a condition assumed to be necessary for an optimum throughput.

— In all the methods discussed up to now it is necessary to fix from the first all parameter combinations for which the dose field and the throughput are to be calculated. An alternative would be to use the results of the calculations with one combination of parameters as a basis to fix the conditions for the next calculation step by the knowledge and experience of the investigator. Such an aim-oriented method, which can be very successful when the structure of the problem is well known, is used, e.g., by *Remer* [6]. (In this example the objective was mainly to increase the field homogeneity, i.e. to decrease $\max (U_i)$.) Also for this method the reduction of the number of parameters by a knowledge of the extrema in the dose field will be useful.

— An effective combination of the stochastic search and an aim-oriented step-by-step procedure is given by the evolution method, as proposed and elaborated mainly by *Bremermann* [7], *Rechenberg* [8], and *Schwefel* [3]. This sequential procedure simulates (above all in its more elaborated variants which are not considered here) biological evolution and proves to be favourable especially for complicated model functions. Therefore it was selected to solve the optimization problem given in eq. (1).

The principle of the evolution method consists in the following steps:

— An admissible starting point P in the parameter space is fixed, i.e. a point that satisfies all constraints. How such a point is found is outlined below.

— Starting from P the next point P' is calculated by a stochastic process. A set of random numbers Z is generated and P' is obtained by

$$P' = P + P \cdot V \cdot Z \quad (3)$$

where

$$V = \sigma / P \quad (4)$$

is the variation coefficient equal for all p_i. The random numbers Z can be distributed uniformly in a given intervall [9], but more suitable are numbers that obey a (0; 1) normal distribution [8].

— It is tested if the dose constraints are satisfied. If not, the calculation is repeated with P again as a starting point. If so, the objective functions calculated with P and P' are compared:

$$L(P') > L(P)\,? \quad (5)$$

If not, start again with P. If so, P' is now the starting point, and the iteration is continued.

Thus a new point is only accepted if it yields a better objective function and at the same time satisfies the dose constraints. The latter requirement corresponds to a barrier method [2]. In which way the parameter constraints (1 c) and (1 d) can be taken into consideration is outlined below.

Implementation of the optimization method

In principle it is not difficult to translate the solution of the optimization problem by means of the evolution method into a computer program. As this procedure is an iterative one considerations are necessary as to the strategy of attenuation in the solution process and as to the criterion for terminating the computation. The procedure starts with a given variation coefficient V to calculate the normally distributed parameter corrections (eq. (3) and (4)). Now all iteration cycles are counted that lead to an maximum overdose $\max(U_i)$ greater than U_{max} (provided a starting point with $\max(U_i) < U_{max}$ is already given). If this number exceeds a certain limit the variation coefficient V is reduced to $(^2/_3) \cdot V$. In this way the correction vector decreases and the search for an optimum is concentrated more and more on an area where $\max(U_i) \approx U_{max}$. The iteration terminates when V, after several reductions, is smaller than a given limit or when a given number of iteration cycles is performed.

Strictly speaking this attenuation and termination strategy does not even guarantee the approximation of a local optimum. It is recommended, therefore, to perform calculations from different starting points. If necessary, such repeated calculations are also useful to select the global optimum among several local ones within the admissible range of the parameters.

The model equations contain a linear parameter with the dimension of time (T, $1/v$, or $1/\omega$) which can be exempted from the general parameter correction. After each iteration step it is determined in such a way that the minimum of all doses D_i calculated by the model equations equals the given value D_{min}. Hence the objective function always assumes the maximum value consistent with the actual parameters.

It remains to be explained how the parameter constraints (1c) and (1d) are taken into consideration. Because of the random parameter corrections they must be observed after each iteration step anew. After the determination of a point P' according to (3) (which, in general, does not satisfy (1c)) a point P'' is calculated which is the nearest one to P' in the parameter space satisfying (1c). This leads to

$$\left.\begin{aligned} &(p_i'' - p_i')^2 \rightarrow \text{Min.} \\ &\sum_l a_{jl} p_l'' - b_j = 0 \end{aligned}\right\} \qquad (6)$$

a least squares problem with constraints which can be solved using *Lagrange*-multiplies; e.g. [10].

If one of the p_i'' exceeds a given limit (constraints (1d)) it is fixed at this limit and all the other corrections are proportionally reduced in order to maintain the direction of the correction vector in the space.

The procedure described up to now requires that the starting point satisfies the dose constraints. But when a problem is more complicated it may be difficult to specify such a point. This problem is solved here such that one can start from any point (which must only satisfy (1c) and (1d) what is ensured easily), that likewise by the evolution method a point is found which guarantees (1b) (i.e. an optimization which aims at a minimum of the maximum U_i) and after the finding of any point of such a kind the optimization is switched over to the actual problem (1a) to (1d). If it is not possible to define an admissible point (which can happen only if the dimensions of the irradiation goods are fixed) the computation stops after the maximum number of iteration cycles. — Of course the minimization of the maximum overdose can also be treated as the main problem without any consideration of the throughput.

The whole method for the solution of (1) was implemented on an ES 1040 computer. The program consists of a problem-independent main part (calculation of the parameter corrections, evaluation of the result of the iteration step, control of the course of the iteration) and of an exchangeable subprogram to calculate the dose in selected field points according to the model equations. The field coordinates x_i^*, y_i^*, z_i^* are not given as fixed values but relative to the dimensions of the irradiation goods (e.g. to s and d in eqs. I/(9), (13), or to A_x, A_y, A_z in I/(15), (18)). In this way always the total volume is covered in the computation of the dose field. Field coordinates can also be fixed relative to source positions. The dose subprogram is called once in each iteration step. For the generation of normally distributed random numbers, for numerical integration and for the solution of the system of linear equations (6) library programs are used (ZUNORM, SIMPS, DMATIN from the ESER library MABIF).

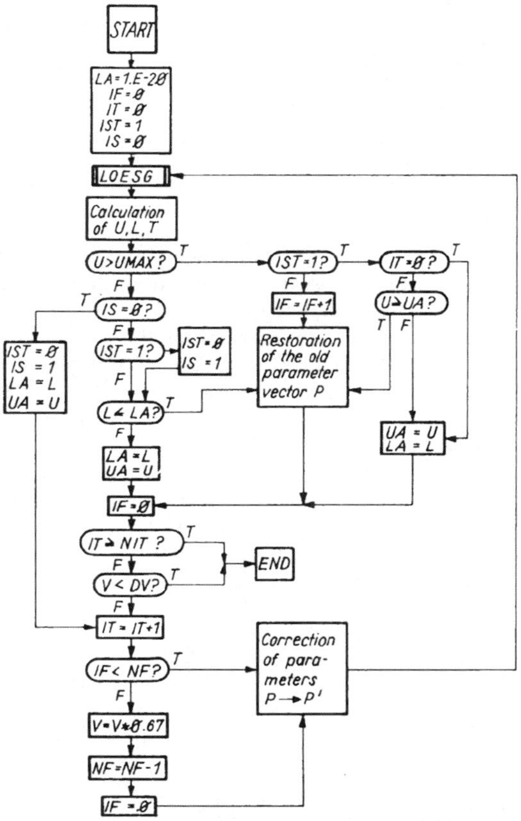

Fig. 1. Flow chart of the main computer program for optimization. The subroutine LOESG represents the exchangeable, model dependent part.

Fig. 1 gives the flow chart of the optimization program. The data input consists of the starting values of the model parameters, the coefficients a, b, and c of the parameter constraints (1c) and (1d), instructions concerning the selection of the field coordinates, quantities indicating the objective of the radiation procedure (DMIN, UMAX), and control parameters (NIT: maximum number of iterations, NF: maximum number of transgressions of the overdose boundary UMAX at a given variation coefficient, V: starting value of the variation coefficient, DV: terminal value for the variation coefficient). The subroutine LOESG represents the exchangeable, model dependent part of the procedure and calculates $D(x^*, y^*, z^*)$ as well as $U = \text{MAX}\,(D(x^*, y^*, z^*))/\text{MIN}\,(D(x^*, y^*, z^*))$, the time parameter ($T$ is taken here as an example), and the throughput L. Correction of parameters means calculating P' according to (3) and taking into account (1c) and (1d). (The old parameter values P are not cancelled because of a possible restoration of the respective initial vector.) The parameters IST and IS serve for the control of the decrease of U (if necessary in the beginning) and for the switching over to the maximization of L. Final results are a set of parameters which yield a throughput LA and a maximum overdose UA.

Examples

The efficiency of the computer method will be demonstrated by two examples. According to the original problem it is supposed that the irradiation goods are onions, i.e. an aqueous medium of density $\varrho = 0{\cdot}5\ g/cm^3$. The material constants used are therefore $\mu = 0{\cdot}063\ cm^{-1}$, $A = 8{\cdot}9$, $\alpha_1 = -0{\cdot}095$, $\alpha_2 = 0{\cdot}06$ for water, and, if needed, $\mu = 0{\cdot}414\ cm^{-1}$, $A = 7{\cdot}4$, $\alpha_1 = -0{\cdot}068$, $\alpha_2 = 0{\cdot}0475$ for iron. The assumed activity is $Q = 370$ TBq with $f = 0{\cdot}01$ in the first and $f = 0{\cdot}0067$ in the second example and a time utilization coefficient $f_z = 1$. The objective of the irradiation was given by $D_{min} = 30$ Gy and $U_{max} = 2{\cdot}33$.

Cubic container

A container of 3 equal dimensions $A_x = A_y = A_z = A$ with walls neglected in the computation is to be irradiated by one source. After the time T, half of the total irradiation time, the container is to change its position according to Fig. 2 (or, which means the same, is to be turned by 180° whithout changing the position). The calculation of the dose field is accomplished by an appropriate subroutine LOESG which was written in a more general form (different container dimensions, 8 positions around the sources) for the development of a new container irradiation facility [11]. The optimization problem reads (for the definition of the two coordinate systems, of (x_s, y_s) and of (x_0, y_0) see part I/Fig. 4 and eqs. (16) ... (18)):

$$L = 8A^3\varrho/T \rightarrow \text{Max.} \tag{7a}$$

$$D_{min} \leqq D_i(x_i^*, y_i^*, z_i^*, h, T) \leqq D_{max} \tag{7b}$$

$$-A \leqq x_i^*, y_i^*, z_i^*, \leqq +A$$

with

$$D_i(x_i^*, y_i^*, z_i^*)$$

$$= T \cdot f \cdot FQ \sum_{k=1}^{2} \frac{1}{d_{ik}^2} \left\{ A e^{-\frac{\mu}{\rho_w} d'_{ik} \rho (1-\alpha_1)} + (1-A)\, e^{-\frac{\mu}{\rho_w} d'_{ik} \rho (1-\alpha_2)} \right\}$$

$$d_{ik}^2 = x_{ik}^2 + y_i^2 + z_i^2$$

$$d'_{ik} = d_{ik} \frac{\sqrt{(x_{s_k} - x_{ik})^2 + (y_{s_k} - y_i)^2}}{\sqrt{x_{ik}^2 + y_i^2}}$$

The coordinates (x^*, y^*, z^*) for D_{min} and D_{max} are known a priori in this problem (there is only one independent position each time):

$$\left.\begin{array}{llll} x^*_{min} = 0 & & x^*_{max} = \pm A \\ y^*_{min} = \pm A & \text{or} \quad 0 & y^*_{max} = 0 \\ z^*_{min} = 0 & \text{or} \quad \pm A & z^*_{max} = 0 \end{array}\right\} \tag{8}$$

Therefore the optimization problem reduces to

$$L = 8A^3\varrho/T \rightarrow \text{Max.} \tag{9a}$$

$$\left.\begin{array}{l} D(x^*_{min}, y^*_{min}, z^*_{min}) = D_{min} \\ D(x^*_{max}, y^*_{max}, z^*_{max}) = D_{max} \end{array}\right\} \tag{9b}$$

Open parameters are A, h, and T. Eqs. (9b) reduce this number to one.

The problem can be solved by substitution of (9b) into (9a), differentiation of (9a) with respect to the remaining parameter h and solution of

$$\partial L(h)/\partial h = 0 \tag{10}$$

The practical problems in each step may be enormous. A more elegant version is the introduction of Lagrange-multipliers. A set of nonlinear algebraic equations have to be solved on a computer, and also the differentiation is best done by numerical methods.

Tab. 1. Solution of the optimization problem of eq. (9) by a pattern search method with a reduced number of parameters. A is given, h and T are calculated according to eq. (9b)

A, cm	h, cm	T, h	L, t/h	max (U)
20	28·2	0·038	0·842	2·333
30	46·6	0·107	1·012	2·333
35	58·0	0·167	1·028	2·333
40	71·2	0·254	1·007	2·333
50	105·4	0·566	0·883	2·333

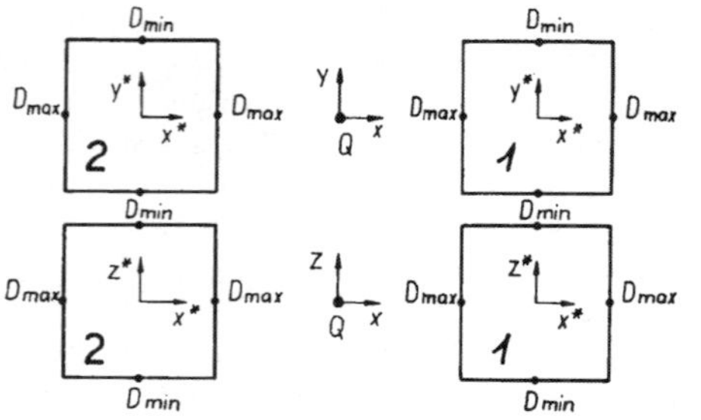

Fig. 2. Irradiation of a container in two positions (intermittent regime) and unchanged orientation in space.

The two equations (9b) render possible a pattern search method with only one parameter. Tab. 1 presents the resulting parameters h and T for given A, as well as the throughput L. Obviously there is only one maximum. For this simple problem the pattern search method is quick and locates the optimum with sufficent exactness. In order to demonstrate the working-method of the evolution procedure several computations were performed:

— Starting values $A = 50$ cm, $h = 60$ cm, $V = 0{\cdot}01$, $NF = 15$. The parameters yield $L = 1{\cdot}38$ t/h and max (U) $= 4{\cdot}47$. After 100 iteration steps max (U) $= 3{\cdot}0$ and $L = 1{,}22$ t/h. Obviously V was taken too small.

— The same starting values, but $V = 0{\cdot}15$. After 14 steps max (U) fell short of the objective value 2·33, and after 150 steps the calculation terminated with $V = 0{\cdot}067$, $A = 38{\cdot}3$ cm, $h = 66{\cdot}5$ cm, $T = 0{\cdot}221$ h, $L = 1{\cdot}017$ t/h, and max (U) $= 2{\cdot}331$. There was the danger that the computation converged too quickly near the parameter point where $U_{max} = 2{\cdot}33$ was reached by chance.

— The next calculation therefore started with parameters that gave max (U) $< 2{\cdot}33$ ($V = 0{\cdot}15$, $A = 30$, $h = 80$, max (U) $= 1{\cdot}49$, $L = 0{\cdot}575$). After 150 iterations: $A = 35{\cdot}6$ cm, $h = 59{\cdot}6$ cm, $T = 0{\cdot}176$ h, max (U) $= 2{\cdot}328$, $L = 1{\cdot}026$ t/h.

— The last calculation started from another point ($V = 0{\cdot}15$, $A = 40$, $h = 76$, max (U) $= 2{\cdot}19$, $L = 0{\cdot}947$). After 150 steps: $A = 34{\cdot}6$ cm, $h = 57{\cdot}0$ cm, $T = 0{\cdot}161$ h, max (U) $= 2{\cdot}33$, $L = 1{\cdot}027$ t/h. There are only insignificant differences from the results of the preceding calculations.

Tab. 2. Radiation doses [Gy] applied to the irradiation goods on a turning table as functions of the field-point coordinates y_i^*, z_i^* [cm]. The parameters used for calculation are the results of the optimization.

y_i^*	z_i^*										
	·65	3·2	6·5	9·7	13·	16·2	19·4	22·7	25·9	29·2	31·8
−30·	52·4	45·4	38·5	33·9	31·2	30·0	30·2	31·9	35·5	41·6	48·2
−24·	67·7	53·5	43·0	37·0	33·7	32·3	32·7	34·9	39·8	49·4	63·6
−23·5	67·8	53·6	43·1	37·1	33·8	32·4	32·7	35·0	39·9	49·5	63·8
−18·	56·9	48·9	41·4	36·6	33·8	32·6	32·8	34·6	38·4	45·0	52·6
−12·	42·6	40·4	37·0	34·3	32·5	31·6	31·6	32·6	34·6	37·6	40·1
− 6·	35·7	35·4	33·8	32·3	31·1	30·5	30·4	31·0	32·1	33·7	34·8
0·	34·1	34·2	33·0	31·7	30·8	30·3	30·2	30·6	31·5	32·9	33·7
6·	36·6	36·2	34·5	32·9	31·7	31·1	31·1	31·8	33·1	34·9	36·1
12·	44·4	42·0	38·2	35·3	33·4	32·6	32·8	34·1	36·5	40·1	43·0
18·	59·4	50·7	42·7	37·6	34·8	33·6	34·1	36·4	40·9	48·7	57·7
23·5	69·7	54·7	43·9	37·7	34·4	33·1	33·8	36·5	42·2	53·2	69·8
24·	69·3	54·5	43·7	37·5	34·2	33·0	33·6	36·3	42·0	53·0	69·4
30·	52·1	45·0	38·1	33·6	31·0	30·0	30·5	32·5	36·6	43·3	50·6

Turning table

The optimization problem for the turning table is given in detail in eq. I/(13) and in Fig. I/3. The assumptions for the calculation are mentioned at the beginning of this chapter. The discrete parameter s is fixed at $s = 30$ cm and R at $R = 170$ cm. These values give a table radius of 200 cm.

Because of the asymmetry of the dose field and the great number of parameters the evolution method is the most appropriate one. Parameters to be determined are d, h_1, h_2, $Q_1 \ldots Q_4$, $z_{Q_1}, \ldots, z_{Q_4}$, $r_{Q_1}, \ldots, r_{Q_4}$; 11 of them are independent. To reduce this number additional parameter constraints are introduced:

$$\left.\begin{aligned} R - r_{Q_1} &= r_{Q_2} - R \\ R - r_{Q_3} &= r_{Q_4} - R \\ r_{Q_1} &= r_{Q_3} \end{aligned}\right\} \qquad (11)$$

As computations have shown these constraints speed up the optimization but have only a minor influence on its result.

The computing time runs up to several hours, and the calculations are made in several stages. Only the final results will be given here:

$$L = 2{\cdot}24 \text{ t/h} \qquad \max(U) = 2{\cdot}33 \qquad \omega = 36{\cdot}16 \text{ h}^{-1}$$
$$d = 32{\cdot}4 \text{ cm} \qquad h_1 = 6{\cdot}0 \text{ cm} \qquad h_2 = 7{\cdot}1 \text{ cm}$$
$$Q_1 = 62{\cdot}6 \text{ TBq} \qquad Q_2 = 95{\cdot}3 \text{ TBq}$$
$$Q_3 = 88{\cdot}9 \text{ TBq} \qquad Q_4 = 123{\cdot}2 \text{ TBq}$$
$$z_{Q_1} = z_{Q_2} = 38{\cdot}4 \text{ cm} \qquad z_{Q_3} = z_{Q_4} = -7{\cdot}1 \text{ cm}$$
$$r_{Q_1} = r_{Q_3} = 146{\cdot}5 \text{ cm} \qquad r_{Q_2} = r_{Q_4} = 193{\cdot}5 \text{ cm}$$

Moreover the resulting dose distribution is of interest. In order to demonstrate which field points were used for the computation the distribution is given in the form of a table (Tab. 2). The coordinates (in cm) are calculated from given fractions of the distances d and $2s$ and from the positions (r_{Q_j}) of the sources. The asymmetry of the dose field is clearly demonstrated.

Conclusion

Given an optimization problem as in eq. (1) it is supposed that a definite global optimum exists. Any algorithm must be judged by its ability to approximate this optimum and by the speed at which this is done. In this context some improvements are still desirable. The assumption indicated by eq. (2) has to be proved. This would allow the attenuation and termination strategy based on it to be extended. The number of field points could be automatically enlarged during the optimization, if necessary, and also the starting from different parameter points could be controlled by the computer. But such improvements cost much computer time so that a checking of intermediate results by the investigator should not be omitted.

A final remark concerns the relative value of mathematical optimization. In the container example of the preceding chapter the third and fourth calculations show that from a starting point which is quite reasonable the throughput can at best be doubled by the optimization. But the design of a facility as in Fig. 5 of part I with all positions occupied at each time allows the final result of the simple example to be improved more than fourfold. As already stated in part I fundamental changes in the design are the main source for a more efficient facility. But the design as well as the control of the irradiation regime will become a more and more complicated task. In order to make the best use of all the efforts an optimum choice of all parameters is a matter of great importance.

Acknowledgements

I would like to thank Dr. habil. *G. Hübner*, Dr. *H.-W. Schulze* and Dr. *M. Remer* (ZfI, Leipzig) as well as Ing. *P. Hargittai* (Institute of Isotopes, Budapest) for advices and many stimulating discussions.

Received April 3, 1985
Accepted in revised form May 9, 1985

References

[1] *E. Winkler*, Isotopenpraxis *22* (1986) 1
[2] *K.-H. Elster*, Nichtlineare Optimierung. Leipzig: B. G. Teubner Verlagsgesellschaft 1976
[3] *H.-P. Schwefel*, Numerische Optimierung von Computer-Modellen mittels der Evolutionsstrategie (Interdisciplinary Systems Research, 26). Basel: Birkhäuser-Verlag 1977
[4] *H.-W. Jank, A. Meister*, Kulturpflanze *30* (1982) 125
[5] *P. Hargittai, V. Stenger*, ZfI-Mitt. *17* (1983) 40
[6] *M. Remer*, ZfI-Mitt. *71* (1983) 47
[7] *H.-J. Bremermann*, Math. Biosci. *9* (1970) 1
[8] *I. Rechenberg*, Evolutionsstrategie. Stuttgart: Fromann-Holzboog 1973
[9] *P. Koeppe, C. Hamann*, Computer Progr. Biomed. *12* (1980) 121
[10] *R. Zurmühl*, Praktische Mathematik für Ingenieure und Physiker. Berlin – Heidelberg – New York: Springer 1965
[11] *G. Hübner*, ZfI-Mitt. *98* (1984) 578

Isotopenpraxis 22 (1986) 1, S. 12 bis 16

Ion-Induced X-Ray Emission: Parameter Set for Attenuation Corrections

G. Brunner

(Zentralinstitut für Isotopen- und Strahlenforschung der AdW der DDR, Leipzig)[1])

Analytical applications of ion-induced X-ray emission demand an interpolation procedure of X-ray attenuation coefficients. For photo attenuation a three-parameter representation is widely used. Its reliability has been tested under the aspects of Scofield's RHS calculations. Based upon these results parameter sets are proposed for the total attenuation coefficients in the interval $1 \leqq E\,[keV] \leqq 40$ which is useful for PIXE.

Eine analytische Anwendung der ionen-induzierten Röntgenstrahlenemission erfordert eine Interpolation der Röntgenstrahlungs-Schwächungskoeffizienten. Für die Photo-Schwächung ist eine drei-parametrische Darstellung weit verbreitet. Ihre Zuverlässigkeit wurde unter den Aspekten der Scofieldschen RHS-Rechnungen geprüft. Auf der Grundlage dieser Ergebnisse wird ein Parametersatz für den Gesamt-Schwächungskoeffizienten vorgeschlagen, der im Bereich von $1 \leqq E\,[keV] \leqq 40$ für PIXE brauchbar ist.

Keywords

attenuation; PIXE analysis; total cross sections; X radiation

Introduction

Analytical application of ion-induced x-ray emission demand an interpolation procedure of x-ray attenuation coefficients [1]. For photo attenuation a three-parameter representation $\tau = CE^{-\alpha}Z^{\beta}$ (E x-ray energy, Z atomic number) is widely used [2]. Its reliability has been tested under the aspects of *Scofield's* RHS calculations [3]. Based upon these results parameter sets are proposed for the total attenuation coefficients $\mu/\varrho = (\tau + \sigma)/\varrho$ in the interval $1 \leqq E$ (keV) $\leqq 40$ which is useful for PIXE. To decide the correct energy branch for calculation, the K-, L- and M-absorption edge energies can be calculated by two further parameters in such a way that Z and the line type are the only inputs to supply μ/ϱ of all elements for an x-ray line.

1. Atomic Photo Cross Sections

Spectra evaluation of thick target PIXE demands a useful interpolation sheme of x-ray attenuation coefficients. The old and well-known representation [4]

$$\frac{\tau}{\varrho} = C_{\lambda} Z^{\beta} \lambda^{\alpha} \tag{1}$$

is suited for the share of photo effect due to its close relation to the formula

$$\frac{\tau}{\varrho} = \frac{4\pi e^2}{mc^2}\frac{L}{A}\lambda^3 \sum_q \frac{z_q}{\lambda_q^2} \tag{2}$$

on the basis of a nonquantified theoretical model (z_q number of electrons in the q level, λ_q wavelength of the q absorption edge, L *Avogadro's* number and A atomic weight) and its first approximation

$$\frac{\tau}{\varrho} \approx \frac{4\pi e^2}{mc^2}\frac{L}{A}\lambda^3 R^2 \frac{z_q}{n_q^4}(Z-S)^4 \tag{3}$$

by substitution of

$$\lambda_q = \frac{n_q^2}{R(Z-S)^2} \tag{4}$$

(R *Rydberg*-constant, S screening constant, n_q main quantum number for the q shell).

According to *Theisen* and *Vollath* [2] a linearisation by taking the logarithm of τ/ϱ in eq. (1) and introduction of the necessary differential condition for minimum square difference between the logs of the available data basis and the interpolation values yield $\ln C$, α and β in a simultaneous minimisation procedure of all $\tau - Z - \lambda$ triples.

In this way *Theisen* and *Vollath* obtained their useful (C, α, β)-parameter sets for energies $1 \leqq E\,[\mathrm{keV}] \leqq 18$ in intervals determined by these limits together with the K, L and M_1 edges for atomic numbers $3 \leqq Z \leqq 94$ in intervals given by the periods of the Periodic System of Elements.

In this paper all main features of the formal *Theisen-Vollath*-procedure have been maintained but

1) the theoretical atomic photo cross sections of *Scofield* [3] constitute the data basis instead of the less mighty, less systematically distributed body of Theisen's experimental attenuation coefficients;
2) the scheme was worked out much more completely and does now include all possible E-Z-combinations within the mentioned interval limits of *Theisen* and *Vollath*;
3) moreover it contains also $Z = 1$ and 2;
4) besides $1 \leqq E\,[\mathrm{keV}] \leqq 20$ the scheme supplies parameters also in $20 \leqq E\,[\mathrm{keV}] \leqq 40$ in order to meet better the general PIXE conditions.

As for as $n°$ 1 is concerned, such a theoretical basis is justified by an extensive comparison of all experimental photo cross sections on the one hand and Scofield's results on the other, done by *Veigele* and *Hubbell* [5], and thus it is much more reliable in providing an answer as to what is the real value of an interpolation formula like eq. (1), than such results which discard values, differing by more than e.g. 25% from a first interpolation run [2], particularly the

[1]) Address: DDR-7050 Leipzig, Permoserstrasse 15

Tab. 1. Parameters C, α and β of eq. (5); $1 \leq E$ [keV] ≤ 20

		$E > E_K$	$E_K > E > E_{L1}$	$E_{L1} > E > E_{L2}$	$E_{L2} > E > E_{L3}$	$E_{L3} > E > E_{M1}$
$1 \leq Z \leq 2$	C	22·3*)				
	α	3·378				
	β	4·338				
$3 \leq Z \leq 10$	C	20·1				
	α	3·070				
	β	4·320				
$11 \leq Z \leq 18$	C	34·5	0·779			
	α	2·814	2·680			
	β	3·882	4·335			
$19 \leq Z \leq 36$	C	65·1	2·76	330	$1 \cdot 12 \times 10^{15}$	0·0188
	α	2·723	2·658	1·922	− 1·679	2·383
	β	3·601	3·900	2·392	− 6·268	4·706
$37 \leq Z \leq 54$	C	135·4	2·16	18·7	65·0	0·0928
	α	2·570	2·687	2·390	2·177	2·408
	β	3·274	3·984	3·274	2·800	4·271
$55 \leq Z \leq 71$	C		3·15	26·9	4·12	0·317
	α		2·646	2·380	2·573	2·462
	β		3·867	3·179+	3·650	3·977
$72 \leq Z \leq 86$	C		4·27	12·6	1·49	0·309
	α		2·588	2·458	2·692	2·510
	β		3·760	3·399	3·953	3·999
$87 \leq Z \leq 94$	C		4·18	7·13	1·44	0·459
	α		2·580	2·512	2·715	2·526
	β		3·758	3·562	3·976	3·918

*) for $Z = 1$: $C = 22 \cdot 3/2$ (half filled K shell)

Tab. 2. Parameters C, α and β of eq. (5); $20 \leq E$ [keV] ≤ 40

		$E > E_K$	$E_K > E > E_{L1}$	$E_{L1} > E > E_{L2}$	$E_{L2} > E > E_{L3}$	$E_{L3} > E > E_{M1}$
$1 \leq Z \leq 2$	C	24·6*)				
	α	3·437				
	β	4·466				
$3 \leq Z \leq 10$	C	24·4				
	α	3·349				
	β	4·568				
$11 \leq Z \leq 18$	C	22·0				
	α	3·117				
	β	4·353				
$19 \leq Z \leq 36$	C	37·4				
	α	2·895				
	β	3·923				
$37 \leq Z \leq 54$	C	78·1	83·2			
	α	2·687	4·214			
	β	3·519	4·220			
$55 \leq Z \leq 71$	C	70·9	1·42			
	α	2·655	2·761			
	β	3·512	4·141			
$72 \leq Z \leq 86$	C		1·73			
	α		2·668			
	β		4·022			
$87 \leq Z \leq 94$	C		2·220	4·56	1·55	
	α		2·590	2·547	2·690	
	β		3·906	3·684	3·942	

*) for $Z = 1$: $C = 24 \cdot 6/2$ (half filled K shell)

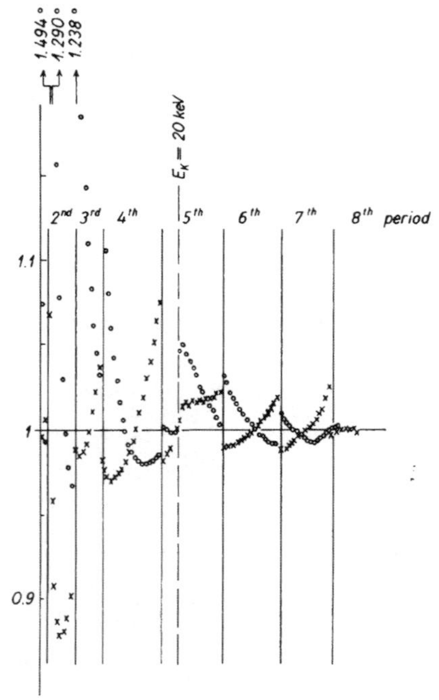

Fig. 1. interpolated τ (this work) divided by τ of *Scofield* [3] versus Z in $1 \leqq Z \leqq 94$ (horizontal axis, lin. scale)

$E = 20$ keV; dots: parameters for $1 \leqq E$ [keV] $\leqq 20$ used, crosses: parameters for $20 \leqq E$ [keV] $\leqq 40$ used

reliability of x-ray attenuation data tends rather generally to be overestimated.

Adapted to the needs of energy dispersive measurements the tables 1 and 2 give parameter sets for C, α (here as energy exponent) and β in

$$\tau[\mathrm{b}] = CE\,[\mathrm{keV}]^{-\alpha}\,Z^{\beta} \tag{5}$$

The effect of the interpolation procedure is worst at the boundaries of the validity intervals. Fig. 1 gives interpolated values according to Tab. 1 and Tab. 2 in units of *Scofield's* results for 20 keV and all elements. Maximum departures from the "right" (Scofield) values up to 50% occur due to the interpolation in the second period for the parameters of the widespread E interval 1 to 20 keV. Such large interpolation errors for low Z of the second and third period are mainly due to the high relative shielding contribution S/Z (see chapter 3), the K-edge height varying strongly with Z and pressed together into more extended Z intervals and to systematic α departures near the edges. As an extreme case the average interpolation error amounts to 14% for 20 keV in the second period, taking the 1 to 20 keV parameter set. Fig. 2 gives more characteristic results for 10 keV: the average interpolation error is now 8% in the second period (Fig. 2a). Fig. 2, parts b to d, gives respective results in comparing interpolated values to some of the most widely used compilations [6, 7, 2]. The *Storm* and *Israel* results [6] (theory similar to *Scofield*: *Hartree-Slater* but somewhat lower exchange contributions and based upon experimental binding energies) agree very well with *Scofield's* values. The experimentally founded data of *Veigele* [7], and even more those of *Theisen* and *Vollath* [2], give about 3% and 7%, respectively, higher amounts. In the latter case the interpolation parameters for the low-Z and high-Z elements seem to be not so good. That there is some systematic discrepancy may be expected for the *Theisen-Vollath* compilation near the high-energy limit (17 to 18 keV): for such energies and e.g. in the case of Li ($Z = 3$) the scattering share, σ, of the experimental attenuation is negligible on no account. Pressing it into the photoionisation-like interpolation scheme must lead to "τ" values which are too high.

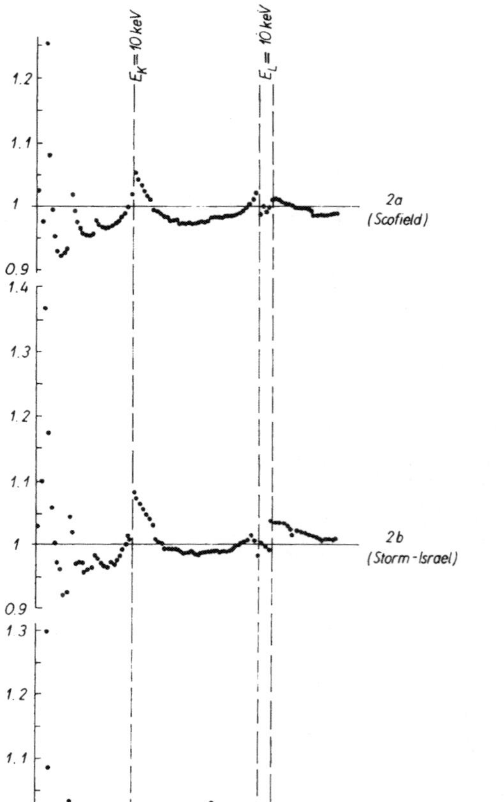

Fig. 2. interpolated τ (this work) divided by τ of some other authors versus Z in $1 \leqq Z \leqq 94$ (horizontal axis, lin. scale)

$E = 10$ keV

2. Total Atomic Attenuation Cross Sections

The scattering contribution depends above all on geometry, therefore it is arbitrary to some extent what σ value should be added to τ to account for scatter. To take total cross sections is standard, and it is justified under PIXE conditions in a wide enough measuring chamber since a) beam spot and effective detector dimensions (mm) may be considered as small with respect to the target detector distance and b) the characteristics of scatter emission for relatively low energies is not so pronounced in forward direction. As far as the really forward acting scatter is concerned, it becomes compensated (or even more than compensated) by a deficiency in the Compton process due to the electron binding noticeable at low angles (incoherent scattering function), so that the Klein-Nishina total collision cross section, σ_{KN}, appears as an appropriate approximation in the atomic value of

$$\sigma_{total} = \tau + Z\sigma_{KN} \quad (6)$$

Since σ_{KN} does not vary from 1 to 40 keV by much more than a dozen percent, an approximation

$$Z\sigma_{KN} \approx Z \times 0{\cdot}6\,[b] \quad [7]$$

seems applicable as far as the total scatter contribution to σ_{total} is not negligible at all. ($Z \times 0{\cdot}66$ [b] marks the limiting case of *Thomson's* scattering model for vanishing energy: ${}_{e}\sigma_{Thom} = \frac{8\pi}{3} r_0^2 \approx 0{\cdot}66$ [b] per electron)

As an example: for Li ($Z = 3$) the total scatter (coherent plus incoherent) reaches 10% of the photo fraction at about 4 keV and surmounts photo interaction by a factor of 126 at 40 keV. The values taken from the simple eq. (7) are 6% and a factor of 133, respectively, which means an error of less than 6% of σ_{total} for both energies in this rather extreme case.

3. Edge Energies

Although a lot of reliable tableworks of edge energies, E_q, exists (see e.g. [8] it is useful to have at one's disposal an equivalent interpolation scheme for computer evaluation of experiments like that of the attenuation coefficients. To decide upon the right C, α, β parameter set to be applied from chapter 1 a higher effort is needed for the edge energies and even line energies as edge energy differences. Therefore it is necessary to distinguish more Z intervals, and they should be chosen by different types of electrons which are inserted in a stepwise buildup of an element from the preceding one in the Periodic Table, since charge penetration and thus shell screening, S, are very different for the different charge types. This screening (mainly a so-called outer screening in contrast to the "inner screening" of *Slater* [9]) describes the departure from a hydrogen-like situation and is of high importance with respect to E_q, mainly for low Z and higher main quantum numbers ($q =$) n.

It has been found useful to define a "reduced screening" $S^x = S/(n^2\sqrt{Z})$ since it is only slightly varying with n and

Tab. 3. Parameters a and b of eq. (9); $E_q > 1$ keV; $R = 13{\cdot}6$ eV assumed; e. g. $\begin{matrix}-1.8670\\(-3)\end{matrix}$ means -1.8670×10^{-3} for a

	n	$Z=$	11–18	19–20	21–30	31–36	37–38	39–46	47–54	55–57	58–70	71–79	80–86	87–94
K	1	a	−1·8670 (−3)	−3·0000 (−3)	−4·8490 (−4)	−3·5143 (−3)	−5·0000 (−3)	−4·7262 (−3)	−7·1429 (−3)	−8·0000 (−3)	−9·5824 (−3)	−1·3567 (−2)	−1·7393 (−2)	−2·100 (−2)
		b	0·6592	0·6800	0·6270	0·7182	0·7720	0·7617	0·8747	0·9200	1·0140	1·2937	1·5976	1·9105
L_1	2	a			2·4500 (−3)	−7·6857 (−4)	−2·1000 (−3)	−1·1679 (−3)	−2·5857 (−3)	−3·1500 (−3)	−2·5973 (−3)	−4·2683 (−3)	−5·9250 (−3)	−7·0357 (−3)
		b			0·4411	0·5381	0·5855	0·5490	0·6157	0·6454	0·6146	0·7324	0·8637	0·9599
L_2	2	a				−1·4786 (−3)	−2·8000 (−3)	−1·6190 (−3)	−2·9310 −(3)	−3·2500 (−3)	−2·7802 (−3)	−4·3817 (−3)	−5·9857 (−3)	−7·1190 (−3)
		b				−0·6147	0·6622	0·6163	0·6780	0·6943	0·6682	0·7812	0·9085	1·0067
L_3	2	a	−5·3964 (−3)	−3·1000 (−3)	3·0406 (−3)	−5·7143 (−4)	−1·3000 (−3)	−2·4762 (−4)	−1·1092 (−3)	−1·1000 (−3)	1·5604 (−4)	−3·2500 (−4)	−9·0000 (−4)	−7·7381 (−4)
		b	0·6562	0 6153	0·4884	0·5963	0·6218	0·5804	0·6210	0·6202	0·5479	0·5825	0·6282	0·6170
M_1	3	a							−1·5112 (−3)	−9·8432 (−4)	3·2685 (−4)	−1·0044 (−3)	−2·1295 (−3)	−2·2702 (−3)
		b							0·4823	0·4529	0·3779	0·4720	0·5613	0·5728
M_2	3	a								−1·4131 (−3)	2·3554 (−4)	2·1714 (−3)	−2·2732 (−3)	−2·3589 (−3)
		b								0·5039	0·4094	0·2600	0·5973	0·6043
M_3	3	a	1·5597 (−3)	−2·4990 (−3)	6·3607 (−3)	−8·4557 (−4)	−2·0693 (−3)	4·5514 (−4)	−7·8811 (−4)	−5·5329 (−4)	1·2151 (−3)	2·0000 (−4)	−6·4992 (−4)	−3·6083 (−4)
		b	0·3383	0·4438	0·2687	0·4798	0·5217	0·4240	0·4840	0·4690	0·3672	0·4393	0·5067	0·4808
M_4	3	a									8·9573 (−4)	−8·8218 (−5)	−9·4331 (−4)	−5·6910 (−4)
		b									0·4376	0·5079	0·5759	0·5428
M_5	3	a				−3·8493 (−3)	−4·0345 (−3)	−2·8662 (−4)	−1·5792 (−3)	−1·7777 (−3)	1·1600 (−3)	2·1851 (−4)	−5·4046 (−4)	−1·8630 (−4)
		b				0·6635	0·6679	0·5221	0·5833	0·5936	0·4259	0·4928	0·5529	0·5214
N_1	4	a											−1·1631 (−3)	−9·5053 (−4)
		b											0·4377	0·4186

Z and thus enables a simpler (because linear) approximation

$$S^x = aZ + b \qquad (8)$$

and hence a highly reliable calculation of binding or edge energy

$$E_q = R\frac{Z^2}{n^2}\left(1 - \frac{n^2}{\sqrt{Z}}[aZ + b]\right)^2 = R\left(\frac{Z - S}{n}\right)^2 \qquad (9)$$

Tab. 3 supplies parameters a and b for all edges from K to N_1 and all elements with those edges above our low-energy limit of 1 keV. The errors are smaller than some 10^{-4}. Maximum departures occur for the 4f elements (rare earth region): $\Delta E_q/E_q \approx 4 \cdot 10^{-4}$ for q = K and $\approx 6 \cdot 10^{-4}$ for q = L_1 in a few special cases.

To supply K_{α_1}, K_{β_1} and L_{α_1} line energies above 1 keV by the differences K-L_3, K-M_3 and L_3-M_5 the parameter scheme has been correspondingly completed.

4. Transformation from Atomic Cross Sections to Attenuation Coefficients

To calculate

$$\frac{\mu}{\varrho} = \frac{L}{A}\sigma_{\text{total}} \approx \frac{1}{A}(\tau + Z \times 0{\cdot}6) \times 0{\cdot}602 \qquad (10)$$

(τ in [b]) a simple representation of the relative atomic weights, A, as a function of Z is needed. Within the framework of all other approximations (a few %)

$$A_i = Z(k_Z A_{i-1}^{2/3} + 2) \quad \text{together with}$$

$$k_Z = 2{\cdot}24 \times 10^{-5} Z + 1{\cdot}334 \times 10^{-2} \qquad (11)$$

(for $Z > 1$, otherwise $A = Z$) has been found as a useful recursive relation based upon the well-known *Bethe-Weizsäcker* formula of nuclear masses, taken from nuclear physics basic literatur (e.g. [10]). For calculations of $A = f(Z)$ at the % level, only very few iterations must be done.

5. Application

All above discussed approximations are fit for a small (micro) computer program in two versions: *version A*: input of the attenuating element, Z_{att}, and of an energy value, leading to the mass attenuation coefficient, μ/ϱ; *version B*: input of Z_{att} and of an emitter element, Z_{em}, together with the choice of the line type (K_{α_1}, K_{β_1}, L_{α_1}) yiels μ/ϱ (plus energies of the lines). Tab. 4 contains results for a version B test run under the special condition $Z_{att} = Z_{em} = Z$ (so called self-attenuation) in order to go across all possible energies and for K_{α_1} in a Z region, for which comparable values of *Theisen* and *Vollath* exist ($1 \leqq E\,[\text{keV}] \leqq 18$). Some further figures from standard tables are included.

Tab. 4. Version B test run: $Z_{att} = Z_{em} = Z$ ("self-attenuation"); results A, E_K, E_{L3}, $EK_{\alpha_1} = E_K - E_{L3}$ and first column of $\mu/\varrho_{EK_{\alpha_1}}$ calculated from the given Z; last three columns are results for $\mu/\varrho_{EK_{\alpha_1}}$ of other authors for comparison; energies in keV

Z	A	E_K	E_{L3}	EK_{α_1}	$\mu/\varrho EK_{\alpha_1}$, cm^2g^{-1}			
					this paper	[2]	[11]	[12]
11	23·22	1·073	0·032	1·041	660	682	561	
12	25·41	1·305	0·049	1·255	479	533	489	
13	27·62	1·560	0·070	1·490	393	426	404	
14	29·84	1·840	0·096	1·744	329	346	356	
15	32·07	2·144	0·126	2·018	280	286		
16	34·31	2·472	0·161	2·311	240	239		
17	36·57	2·825	0·201	2·624	209	203		
18	38·84	3·203	0·246	2·956	183	173		
19	41·12	3·606	0·294	3·312	163	173		
20	43·41	4·036	0·347	3·689	141	150		
21	45·71	4·492	0·402	4·089	124	131		
22	48·02	4·966	0·456	4·510	109	115		111
23	50·34	5·465	0·513	4·952	96·2	102		97·3
24	52·68	5·989	0·575	5·415	85·6	90·5		85·7
25	55·03	6·538	0·639	5·899	76·6	80·8		75·6
26	57·39	7·112	0·708	6·404	68·8	72·5		70·4
27	59·75	7·710	0·780	6·930	62·1	65·4		62·9
28	62·14	8·334	0·855	7·478	56·2	59·2		59·8
29	64·53	8·982	0·935	8·047	51·1	53·8		51·5
30	66·93	9·656	1·018	8·638	46·6	49·0		49·2
31	69·34	10·366	1·1161	9·250	42·6	44·8		
32	71·77	11·104	1·217	9·886	39·1	41·1		
33	74·20	11·868	1·324	10·544	35·9	37·8		
34	76·65	12·659	1·435	11·224	33·1	34·8		
35	79·11	13·476	1·552	11·925	30·6	32·2		
36	81·58	14·321	1·673	12·648	28·3	29·8		
37	84·06	15·198	1·805	13·393	25·8	27·7		
38	86·55	16·105	1·940	14·165	24·0	25·7		
39	89·05	17·037	2·082	14·956	22·4	22·2		
40	91·57	17·998	2·223	15·775	20·9	22·3		
41	94·09	18·986	2·369	16·617	19·5	20·9		
42	96·63	20·002	2·520	17·482	18·2	19·5		

The system works useful in such applications in which many elements and/or energies must be handled.

Received April 22, 1985

References

[1] *J. L. Campbell, J. A. Cookson, H. Paul*, Nucl. Instr. and Meth. *212* (1983) 427

[2] *R. Theisen, D. Vollath* in "Ion Beam Handbook for Material Analysis" (eds. J. W. Mayer and E. Rimini), Academic Press 1977, pp. 450–482

[3] *J. H. Scofield* UCRL-51326 (1973)

[4] *K. Grosskurth*, Ann. Phys. *20* (1934) 197

[5] *J. H. Hubbell, Wm. J. Veigele*, NBS Technical Note 901 (1976)

[6] *E. Storm, H. I. Israel*, Nucl. Data A7 (1970) 565

[7] *Wm. J. Veigele*, At. Data *5* (1973) 51

[8] *J. A. Bearden, A. F. Burr*, Rev. Mod. Phys. *39* (1967) 125

[9] *J. C. Slater*, Phys. Rev. *36* (1930) 57

[10] *T. Mayer-Kuckuk*, „Physik der Atomkerne". Stuttgart: Teubner 1974, p. 45

[11] *B. L. Henke, E. S. Ebisu*, Adv. in X-Ray Ana. *17* (1975) 213

[12] *W. H. McMaster, N. K. Del Grande, J. H. Mallett, J. H. Hubbell*, UCRL-50174 sections I–IV (1969, 1970); siehe auch "Handbook of Chemistry and Physics", Chem. Rubber Publ. Co., Ohio 1974 (55th ed.) E-139 ff.

Isotopenpraxis 22 (1986) 1, S. 17 bis 20

Effects of Gamma Radiation Doses on the Infrared Absorption Spectra of Solid Transition Metal Complexes

M. M. Abou-Sekkina[1]), *M. Y. Elsheikh*

(Chemistry Department, Faculty of Science, Tanta University, Tanta, Egypt

Two organic ligands containing pyridine ring and three of their cobalt and nickel complexes were prepared and recrystallized from their appropriate solvents. Careful courses of infrared absorption measurements before and after various gamma radiation doses (10^5—10^9 rd), elemental analysis and melting point determination were conducted. Materials investigated were arranged in a descending order of their decreasing stability towards gamma irradiation. Results were further discussed in terms of bond-strength and the steric effect of molecular groups.

Von zwei organischen pyridinring-enthaltenden Liganden wurden drei entsprechende Co- und Ni-Komplexe synthetisiert und die IR-Spektren aufgenommen. Nach einer γ-Bestrahlungsdosis von 10^5 ... 10^9 rd wurden neben den Elementaranalysen und Schmelzpunkten auch die Veränderungen in den IR-Spektren untersucht und danach eine Rangfolge mit geringer werdender Stabilität gegenüber der γ-Strahlung aufgestellt. Die Ergebnisse wurden in bezug auf Bindungsstärken und der Auswirkung sterischer Effekte diskutiert.

Keywords

gamma radiation; infrared spectra; transition metal complexes

Introduction

The infrared absorption spectra of the carbonyl oxygen was studied by *Kagarise* [1]. The tendency of the carbonyl oxygen to attract electrons (C^+-O^-) results in an lowering of the C = 0 frequency. *Bellamy* [2] found that the amides have progressively decreasing carbonyl frequencies resulting from the mesomeric effect ($O^--C{=}N^+$), where the oxygen atom draws electron and the nitrogen atom donates electrons. Thus, the stretching vibration frequencies [3, 4] appear in the region 1700 to 1640 cm^{-1}.

The NH stretching vibration [5] gives a weak band at 3500 — 3300 cm^{-1}. A stronger band at 750 — 700 cm^{-1} in secondary aliphatic amines is due to NH wagging [6, 7]. Solution of N-methyl anilines [8] absorb at 3460 — 3420 cm^{-1}.

$^{\nu}$C=C stretching vibration [9, 10] gives rise to a band in the region 1680 — 1620 cm^{-1}.

Gill et al. [11] studied the difference between the spectra of pyridine and coordinated pyridine, showing that the coordinated pyridine is distinguished from free base by the presence of weak band between 1235 to 1250 cm^{-1} and by shift in the strong 1578 cm^{-1} band to 1600 cm^{-1}.

The spectrum [12, 13] of the parent 2-aminopyridine and 2,6-diaminopyridine showed a series of bands located at 1590 cm^{-1} with shoulders at 1520, 1510 and 1620 cm^{-1}, while the stretching frequencies of C=N and C=C of the aromatic nucleus are overlapped. The ν_{NH} stretching frequency appeared at the region 3200 — 3350 cm^{-1}.

Bajue et al. [14] studied the infrared absorption spectra of some Co(III) complexes of 2-nitrose-1-naphthol-4-sulphonate and 2-nitrose-1-naphthol ligands. The intense M-L charge transfer bands in the spectrum of each complex are disappeared.

Nonoyama [15] studied the infrared absorption spectra of Co(II), Ni(II) and Cu(II) complexes of N-(2-picolyl)-2-pyridylacetamide (ppaH). It was found that the amide group of (ppaH) is coordinated through its oxygen atom in complexes such as [Ni(ppaH)$_2$X$_2$ · n H_2O], (X = Cl, Br and SCN) where (ppaH) acts as an N, O bidentate ligand.

The damaging effects of J-radiation on plastics and elastomers had been studied by *Harrington* [16]. The effect of gamma radiation exposure on the spectral properties of various materials had been investigated [17, 18]. Interrupted photo-exposure [19] doses have been found to possess more damaging effect than that of the contineous one.

Hanafi et al. [20] on studying the effect of gamma irradiation on the infrared absorption spectra of indium oxide that irradiation leads to predominantly donor type defects. After irradiation, with a dose of $8{\cdot}6 \times 10^9$ rd the broadening and intensity of the IR band change.

Recently, *Abou Sekkina* and *Shakra* [21] on investigating the effect of gamma irradiation on the electrical and optical properties of some azodyes found that γ-irradiation caused a displacement and broadening of the bands in the visible absorption spectra.

As far as the authors are a were no mention was found in the literature concerning the effect of this ionizing radiation on the prepared organic ligands and their transition metal complexes.

Thus, the purpose of the present investigation is to detect this effect on the prepared materials through studying their infrared absorption spectra before and after a wide range of γ-radiation doses for their recent interest to be used in semiconducting devices.

In order to detect if any liable reversible changes are present, measurements were taken after removing the material from the radiation field with six months.

[1]) Author for correspondence to be addressed

Experimental

A) *Material preparation*

This involves preparation of the organic ligands and their corresponding transition metal complexes. The two organic ligands of 2,6-N,N-dibenzoyl-diaminopyridine (dbapH) and 2,6-N,N-diacetyldiaminopyridine (daapH) were prepared by the usual chemical method and crystallized from benzene-alcohol mixture and water respectively. The three transition metal complexes of $Co(daapH)_2Cl_2$; $Ni(dbapH)_2Cl_2$ and $Ni(daapH)_2Cl_2$ were prepared by the following procedure:

A solution of 0·02 mole of the organic ligand in ethanol was added to the solution of the corresponding metal chloride (0·01 mole) in ethanol with stirring. The obtained crystalline precipitate were decanted, filtered off, washed with ethanol and kept in a vacuum desiccator over P_2O_5 for 48 hours. Following the preparation a careful course of chemical analysis and melting point determination were performed to ensure the correct preparation.

B) *Infrared absorption spectra measurements*

The solid technique using KBr disk was applied. The sample concentration was about 3 mg of the material will mixed with 300 mg of KBr (spec. pure) for 15 min. From these mixtures transparent windows were made using pressures amounting to 220 k · gm/cm².

The measurements were taken at room temperature and repeated more than one time to ensure good repeatability.

C) *Gamma radiation exposure*

For irradiation studies, the samples were irradiated using a ^{60}Co gamma cell (250 Curie at the National Institute of Standard, NIS) at a dose rate of $1{\cdot}2 \times 10^4$ rd/h. The materials were exposed to doses ranging from $1{\cdot}0 \times 10^5$ up to $1{\cdot}0 \times 10^9$ rd at room temperature. The infrared measurements were soonly taken within 15 min after removing the samples from the radiation field.

Results and discussion

The infrared absorption spectra of the prepared organic ligands (dbapH; daapH) as well as their prepared transition metal complexes of $Co(daapH)_2Cl_2$; $Ni(dbapH)_2Cl_2$ and $Ni(daapH)_2Cl_2$ were recorded in Fig. 1 and 2. The main characteristic bands were assigned by comparing their spectra with these of 2,6-diaminopyridine (see Tab. 1 and 2).

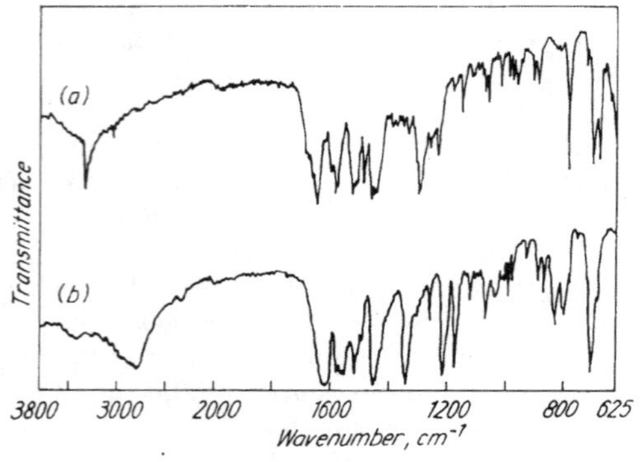

Fig. 1. The obtained room temperature infrared absorption spectra of (a) – dbapH organic ligand and (b) – $Ni(dbapH)_2Cl_2$ metal complex before gamma-irradiation

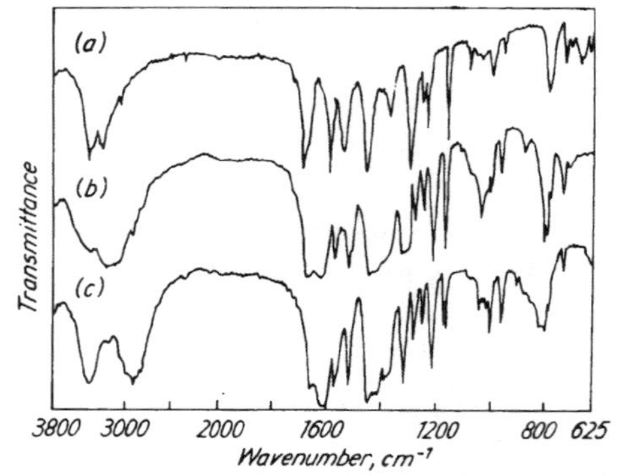

Fig. 2. The obtained room temperature infrared absorption spectra of (a) – daapH organic ligand, (b) – $Co(daapH)_2Cl_2$ and (c) – $Ni(daapH)_2Cl_2$ metal complexes before gamma-irradiation

Tab. 1. The obtained infrared absorption bands of (dbapH) organic ligand and $Ni(dbapH)_2Cl_2$ complex, all cm⁻¹

Material	$\nu_{C=O}$	$\nu_{C=N}$	$\nu_{C=C}$	ν_{NH}
dbapH	1660	1605	1585	3020
$Ni(dbapH)_2Cl_2$	1645	1620	1580	2800 – 3000

Tab. 2. The obtained infrared absorption bands of (daapH) organic ligand; $Co(daapH)_2Cl_2$ and $Ni(daapH)_2Cl_2$ complexes, all cm⁻¹

Material	$\nu_{C=O}$	$\nu_{C=N}$	$\nu_{C=C}$	ν_{NH}
daapH	1690	1615	1580	3040
$Co(daapH)_2Cl_2$	1650	1630	1570	3000 – 3400
$Ni(daapH)_2Cl_2$	1670	1630	1580	2800 – 3100

The spectra exhibit the bands due to ν_{NH} at 2800, 3000 and 3400 cm⁻¹.

After complexation, the band of the amide group is shifted to lower frequency ~20 cm⁻¹ and in some cases ~60 cm⁻¹. But the bands of $\nu_{C=N}$ is shifted to higher frequency ~20 cm⁻¹ and broadened as compared with the parent ligand. This may be attributed to that each of these groups are directly attached to the central metal ion in the different metal complexes. The C=0 stretching frequency appears at 1690 cm⁻¹ and 1660 cm⁻¹ for daapH and dbapH respectively and was found to split on chelation with slight shift of one of the splitted band towards a larger extent or may confused with that due to $\nu_{C=N}$. These phenomena may be correlated to that the ligands acts as a neutral bidentate through only one oxygen atom of the two side chains and the nitrogen atom of the pyridine ring.

The effect of γ-irradiation on the obtained spectra of dbapH and daapH organic ligands as well as those of their metal complexes of $Co(daapH)_2Cl_2$, $Ni(dbapH)_2Cl_2$ and $Ni(daapH)_2Cl_2$ were recorded as a function of γ-irradiation doses of 10^5 up to 10^9 rd in Fig. 3 – 7 respectively. In order to detect the existence of any liable reversible changes of the effect of γ-irradiation, the spectra were further recorded after six months from the last radiation dose [see Fig. 3 – 7, curve (e)].

In irradiation with ^{60}Co (γ-rays) the *Compton* effect has the largest cross section [22] except for materials of very high atomic number, and moreover, the number of atoms displaced ($cm^{-3}\,s^{-1}$) is a maximum in the very light elements and diminishes to zero around atomic weight 125. Thus, this mechanism appears to be more liable for the investigated samples. Irradiation with these rays therefore includes atomic displacement by the *Compton* electrons.

In the following, the effect of gamma irradiation on the prepared materials is given:

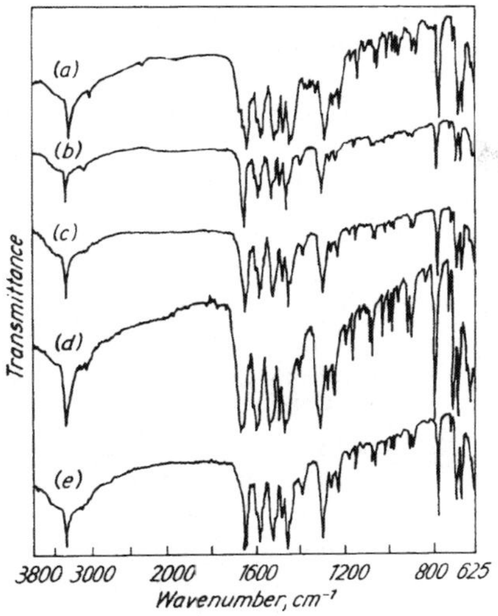

Fig. 3. The obtained infrared absorption spectra of dbapH organic ligand as a function of gamma radiation doses

(a) – before irradiation, (b) – after first dose (10^5 rd), (c) – after second dose (10^7 rd), (d) – after third dose (10^9 rd) and (e) – after six months

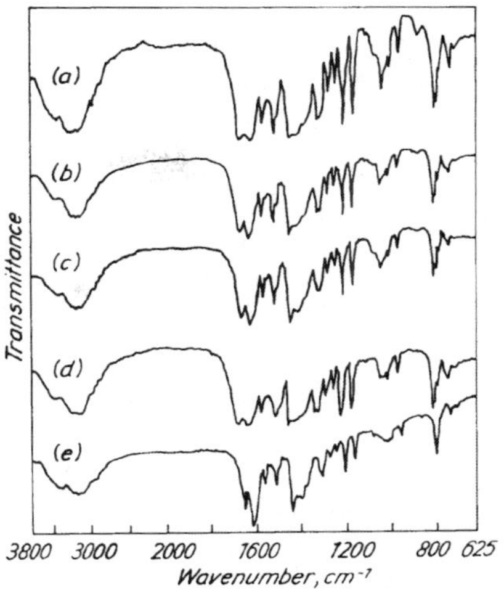

Fig. 5. The obtained infrared absorption spectra of $Co(daapH)_2Cl_2$ as a function of gamma-radiation doses

(a) – before gamma-irradiation, (b) after first dose (10^5 rd), (c) – after second dose (10^7 rd), (d) – after third dose (10^9 rd) and (e) – after six months

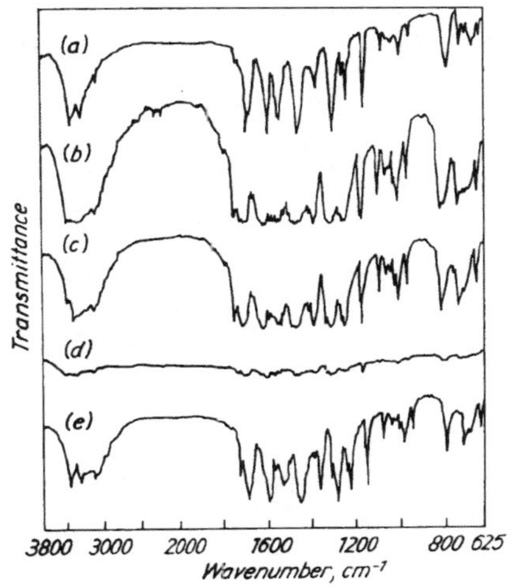

Fig. 4. The obtained infrared absorption spectra of daapH as a function of gamma-radiation doses

(a) – before gamma-irradiation, (b) – after first dose (10^5 rd), (c) – after second dose (10^7 rd), (d) – after third dose (10^9 rd) and (e) – after six months

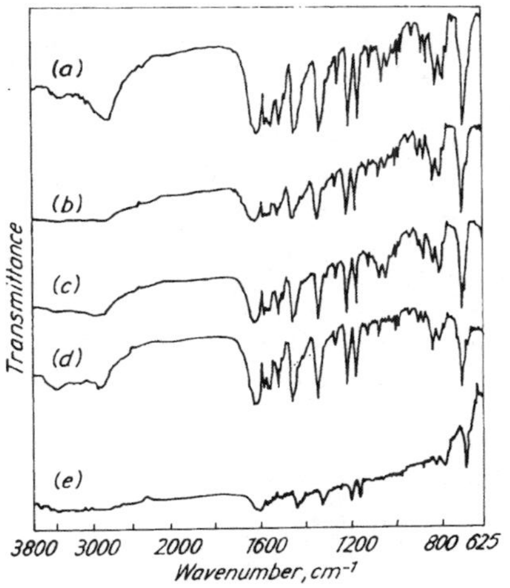

Fig. 6 The obtained infrared absorption spectra of $Ni(dbapH)_2Cl_2$ as a function of gamma-irradiation

(a) – before gamma irradiation. (b) – after first dose (10^5 rd), (c) – after second dose (10^7 rd), (d) – after third dose (10^9 rd) and (e) – after six months

Fig. 3 shows the effect of γ-irradiation doses on the infrared absorption spectra of dbapH organic ligand.

The absorption bands shift towards lower wavelength i.e. higher energy, change in intensity and broadening up to the third irradiation dose (10^9 rd).

Fig. 4 represents the effect of γ-irradiation doses on the infrared absorption spectra of daapH organic ligand.

The absorption bands shift towards shorter wavelength i.o. higher energy, change in intensity and broadening up to the third irradiation dose (10^9 rd). But the effect of -γ irradiation in this case is of less extent than in the forementioned case of dbapH (Fig. 3).

Fig. 5 includes the effect of γ-irradiation doses on the infrared absorption spectra of $Co(daapH)_2Cl_2$ metal complex.

The absorption bands shift towards shorter wavelength (higher wave number), change in broadening and intensity up to the third dose but of greater extent than for the forementioned case of daapH (Fig. 4).

Fig. 6 shows the effect of γ-radiation doses on the infrared absorption spectra of $Ni(dbapH)_2Cl_2$ metal complex.

The absorption bands are displaced towards shorter wavelength, change in intensity and broadening largly at the first dose. At higher doses, the effect becomes less pronounced.

Fig. 7 explain this effect on the infrared absorption spectra of $Ni(daapH)_2Cl_2$ complex.

The absorption bands shift towards shorter wavelength, change in broadening and intensity up to the third dose, but of greater extent than for the forementioned cases (Fig. 3–6).

Accordingly, this may lead us to arrange this group of materials in a descanding order of decreasing their resistance and hence stability towards γ-irradiation doses as follows: daapH > dbapH > $Ni(dbapH)_2Cl_2$ > $Co(daapH)_2Cl_2$ > $Ni(daapH)_2Cl_2$ This depends on the relative weak coordinate chelate bonds in the metal complexes than the covalent bond in the organic ligands,

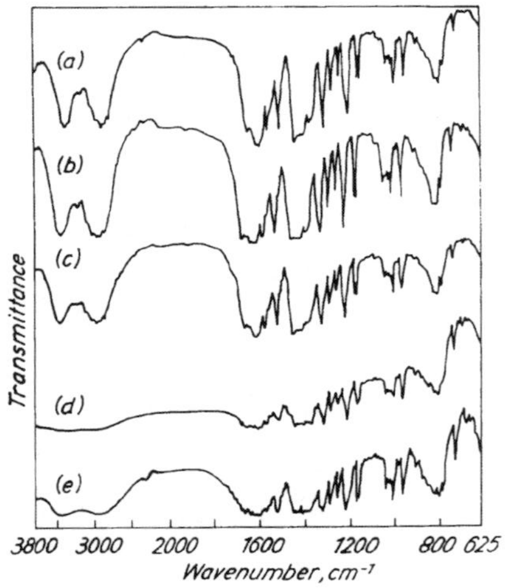

Fig. 7. The obtained infrared absorption spectra of $Ni(daapH)_2Cl_2$ as a function of gamma radiation doses

(a) – before gamma irradiation, (b) – after first dose (10^5 rd), (c) – after second dose (10^7 rd), (d) – after third dose (10^9 rd) and (e) – after six months

The second cause, may be due to the steric effect that existed in the metal complex molecule which makes it less stable and more liable towards the effect of γ-irradiation than for the case of the organic ligand molecule.

In general, the observed shift and broadening of the infrared absorption bands as a result of γ-irradiation may be due to the increase in the concentration of lattice defects [23] for samples of dbapH as well as of daapH and further progressed for the various investigated transition metal complexes. A case which is mostly correlated with the displacement effects [24] which occur when atoms are ejected from their lattice sites by energetic γ-rays. This may be further discussed in the light of the fact that the effect of γ-irradiation on semiconductors; especially for the prepared metal complexes, have significance in displacing atoms from equilibrium position, after sometimes many vacancies and interstitial atoms are formed [25].

The broadening in the infrared absorption spectra, as a result of γ-irradiation is a good indication to the induced lattice imperfection causing a decreased degree of crystallinity. This notation was in conformity with that previously depicted by *Eisen* [26] who stated that damage produced by irradiation with high energy of γ-ray is expected to consist mainly of randomly located interstitial-vacancy spacings which depends on the energy of γ-radiation.

In all cases, measurements of the infrared absorption spectra of the different samples after six months from the last irradiation doses indicate a reversible change. But mostly the absorption bands are still little displaced towards longer wave number (higher energy). This is due to that over the induction period, the point defect are anneal out and the energy released at the site of recombination results in the formation of decomposition product [27] in addition to those induced upon irradiation. Subsequently, the centres of decomposed material grow and creat sufficient strain. Thus, the radiation effect seems to release in each case on standing six months after the final radiation dose due to radiation decay leaving behind a decomposition centres which are responsible for the still observed slight broadening and displacements in the position of the absorption bands.

Conclusion

Accordingly, the effect of γ-irradiation on the prepared materials is a reversible change in the investigated range indicating the disappearance of any destructive effect. This in turn may lead to a conclusion that γ-irradiation up to the third dose (10^9 rd) creats some lattice imperfection with the formation of free radicals, donor centres and/or probably enhance polymerisation through crosslinking. These changes are of great practical use to modifye the semiconducting properties of the test materials and hence a best quality would be obtained for their application in semiconducting devices like transistor and radaars through their interrupted exposure to γ-irradiation doses within the selected investigated range.

Received January 22, 1985
Accepted in revised form March 14, 1985

References

[1] *R. E. Kagarise*, J. Am. Chem. Soc. *77* (1955) 1377
[2] *L. J. Belamy*, Spectrochim. Acta *13* (1958) 60
[3] *R. N. Jones, W. F. Forbes, W. A. Mueller*, Can. J. Chem. *35* (1957) 504
[4] *R. N. Jones E. Spinner*, Can. J. Chem. *36* (1958) 1020
[5] *L. J. Bellamy, K. L. Williams*, Spectrochim. Acta *9* (1957) 341
[6] *J. R. Stewart*, J. Chem. Phys. *30* (1959) 1259
[7] *A. R. Ketritzky, R. A. Jones*, J. Chem. Soc. *1959*, 3674
[8] *P. J. Krueger, H. W. Thompson*, Proc. Roy. Soc. (London) *243* (1957) 143
[9] *S. E. Wiberley, S. C. Bunce, W. H. Bauer*, Anal. Chem. *32* (1960) 217
[10] *H. L. McMurry, V. Thornton*, Anal. Chem. *24* (1952) 318
[11] *N. S. Gill, R. H. Nutall, D. E. Seife, D. W. A. Sharp* J. Inorganic. Nucl. Chem. *18* (1961) 79
[12] *H. Shindo, N. Ikekawa*, Pharm. Bull. (Tokyo) *4* (1956) 192
[13] *H. Shindo*, Ibid *5* (1957) 472
[14] *S. A. Bajue, G. C. Lalar, K. L. Stuart*, J. Inorg. Nucl. Chem. *37* (1975) 51
[15] *M. Nonoyama*, Ibid *37* (1975) 59
[16] *R. Harrington*, Nucleonics *14* (1956) No. 9, 70
[17] *F. M. Ismail, E. M. H. Ibrahim*, Bull N. R. C. *3* (1977) 27
[18] *A. Bishay, A. Kinawi*, Proc. Intern. Conf. Physics of Non Crystalline Solids, North Holland Pub. Co. Amsterdam 1956
[19] *H. Hunter*, J. S.D.C. *56* (1940) 64
[20] *Z. M. Hanafi, E. M. H. Ibrahim, F. M. Ismail* Egypt J. Phys. *8* (1977) 143
[21] *M. M. Abou Sekkina, S. Shakre* (in press)
[22] *R. D. E. Evans*, The Atomic Nucleus McGraw-Hill, New York 1967
[23] *Z. M. Hanafi, F. M. Ismail*, Zeit. Phys. Chem. (N.F.) *1972*, 263
[24] *P. W. Lovy*, Misc. Publ. Nat. Bur. Stand. (U.S.A.) No. 7–12, 21–32 (1966)
[25] *R. D. Dugiahe*, Report of Bristol Conf. on Defects in Crystalline Solids (The Phys. Soc., London) 1952
[26] *F. H. Eisen*, Phys. Rev. *123* (1961) 736
[27] *E. C. Prout, P. J. Harley*, J. Phys. Chem. *66* (1962) 961

Mitteilungen / Information

KERNENERGIE UND WALDSCHAEDEN

In der oeffentlichen Diskussion ueber einen angeblichen Zusammenhang zwischen Kernkraftwerken und Waldschaeden hat sich jetzt auch der Fachverband fuer Strahlenschutz zu Wort gemeldet. Die Wirkung der Strahlung auf Pflanzen sowie ihre Wechselwirkung mit Luftschadstoffen sei der Fachwelt bekannt. Schaedliche Effekte seien dabei allerdings erst bei Strahlenbelastungen festgestellt worden, die erheblich ueber der mittleren natuerlichen Strahlenbelastung in der BRD und der Schweiz liegen (die Fachleute des Strahlenschutzverbandes stammen aus diesen beiden Laendern). Die geringe zusaetzliche Belastung der Umgebung durch Abgabe radioaktiver Stoffe aus Kernkraftwerken sei im Vergleich zur natuerlichen Belastung und Belastung aus dem Fallout der Kernwaffen-Versuchsexplosionen unbedeutend und koenne weder direkt noch indirekt (strahlenchemische Reaktionen auf Luftschadstoffe) beim Waldsterben eine Rolle spielen. (Atomwirtschaft/Atomtechnik 30 (1985) 8/9, S. K+U IV)

Sowjetunion: ERSTE IAEA-INSPEKTION

Aufgrund des Kontrollabkommens fuer sowjetische Kernanlagen, das am 21.2.85 unterzeichnet wurde, haben Inspektoren der IAEA Mitte August 1985 erstmals zwei sowjetische kerntechnische Anlagen inspiziert. Sie wurden aus einer Liste von Anlagen ausgewaehlt, welche die Sowjetunion uebermittelt hatte. Der Kernmaterialkontrolle wurde das 1000-MWe-Kernkraftwerk Nowo-Woronesch-5, Prototyp fuer den WWER-1000, sowie ein 8-MW-Forschungsreaktor in der Naehe von Moskau unterzogen. (Atomwirtschaft/Atomtechnik 30 (1985) 8/9, S. 413)

IAEA: TECHNISCHE ZUSAMMENARBEIT 1984

Die IAEA berichtet, dass 1984 Mittel im Wert von 35,9 Mio.$ zur Verfuegung gestellt wurden, um die technische Zusammenarbeit der IAEA zu unterstuetzen. Die Mittel lagen damit um 1,4 Mio.$ hoeher als im Vorjahr. Die eigentlichen Ausgaben lagen mit 32,6 Mio.$ um 22% ueber dem Jahre 1983. Von den gesamten Mitteln lag der Anteil fuer Nuklearphysik bei 8,2%, fuer Schuerfarbeiten, Abbau und Verarbeitung von Kernmaterial bei 15,9% und der fuer den Bereich Strahlenschutz bei 14,6%. Weitere bedeutende Taetigkeitsgebiete waren die Kerntechnik und -technologie, Landwirtschaft, Medizin, Industrie und Hydrologie. (Atomwirtschaft/Atomtechnik 30 (1985) 8/9, S. 407)

Kurznachrichten / Newlets

Strahlentechnik:
WACHSENDE STRAHLENANWENDUNG

Die IAEA hat einen Internationalen Dienst fuer Dosisvereinheitlichung (IDAS) ins Leben gerufen mit dem Zweck, die Strahlenanwendung in der Industrie weltweit zu verbessern. Die Anwendung grosser industrieller Strahlenquellen waechst mit einer Rate von ca. 25% pro Jahr. Mehr als 250 Bestrahlungsanlagen, in denen hohe Strahlendosen angewendet werden, stehen zur Zeit in 51 Laendern (29 davon Entwicklungslaender) in Betrieb. Der Gesamtwert bestrahlter Produkte wird mit ca. 2 Mrd. Dollar/Jahr vermutlich unterschaetzt. Bekannte Strahlenanwendungen sind Strahlensterilisierung medizinischer Praeparate, Lebensmittelbestrahlung sowie die Bestrahlung von Kunststoffen und anderer Produkte, deren Gebrauch in der modernen Gesellschaft weit verbreitet ist. Ziel des neu gegruendeten Dienstes ist die Standardisierung der Dosimetriemessung, die fuer die Strahlenbehandlung grosse Bedeutung besitzt. (Atomwirtschaft/Atomtechnik 30 (1985) 8/9, S. KuU IV)

Dosimetrie:
STANDARDISIERUNG DURCH IDAS

Eine verlaessliche Dosimetrie ist eine Grundbedingung fuer die technologische Entwicklung in der Strahlenbehandlungsindustrie. Bisher wurden keine gemeinsamen internationalen Anstrengungen unternommen, um Standardisierung der Dosimetriemessung und Dosisvereinheitlichung fuer grosse Strahlenquellen zu erreichen. Diese soll auch die behoerdliche Zulassung und die internationale Freigabe bestrahlter Produkte erleichtern. IDAS soll in diesem Sinne die Qualitaets- und Sicherheitskontrolle foerdern. Bisher haben Aegypten, Argentinien und Belgien um Teilnahme an IDAS offiziell nachgesucht. Betriebsleiter grosser Bestrahlungsanlagen aus zwoelf weiteren Laendern, haben ihr Interesse an einer Teilnahme bekundet. (Atomwirtschaft/Atomtechnik 30 (1985) 8/9, S. 412)

Umwelt:
KERNKRAFTWERKE UND WALDSCHAEDEN

Der Fachverband fuer Strahlenschutz (FS), in dem ueber 700 Strahlenschutzfachleute der BRD und der Schweiz vereinigt sind, hat in einem an die Deutsche Presse-Agentur (dpa) gerichteten Schreiben auf eine Reihe von Tatsachen in Zusammenhang mit der Behauptung hingewiesen, dass auch Kernenergieanlagen an den Waldschaeden beteiligt seien. Die Wirkung radioaktiver Stoffe und ihrer Strahlung auf Pflanzen sowie ihre Wechselwirkung mit Luftschadstoffen sei der Fachwelt schon lange bekannt. Schaedliche Effekte seien dabei allerdings erst bei Strahlenbelastungen festgestellt worden, die erheblich ueber der mittleren natuerlichen Strahlenbelastung in der BRD und der Schweiz liegen. Die geringe zusaetzliche Belastung der Umgebung durch Abgabe radioaktiver Stoffe aus Kernkraftwerken sei im Vergleich zur natuerlichen Belastung und zur Belastung aus dem Fallout der Kernwaffenversuche unbedeutend und koenne schon deshalb weder direkt noch ueber strahlenchemische Reaktionen auf Luftschadstoffe beim "Waldsterben" eine Rolle spielen. Die mehrjaehrige gezielte Ueberwachung des franzoesischen Kernkraftwerks Fessenheim auf deutscher Seite habe als absoluten kurzzeitigen Spitzenwert 39 pCi/m^3 des als Leitnuklid fuer Kernkraftwerke allgemein anerkannten Xe-133 ergeben. Das natuerlich radioaktive Edelgas Ra-222 schwankte jedoch in dieser Region bereits zwischen 120 und 400 pCi/m^3. (Atomwirtschaft/Atomtechnik 30 (1985) 8/9, S. 412

Italien:
TEUERE KERNENERGIE-VERSAEUMNISSE

Die starke Abhaengigkeit von Energieeinfuhren belastet die Aussenhandelsbilanz Italiens hoeher als in anderen Laendern. In einer Erklaerung dazu wies der Praesident des staatlichen Energiekonzerns ENI, F. Reviglio, darauf hin, dass ein rechtzeitiger Ausbau der Kernenergie in den vergangenen Jahren der italienischen Wirtschaft ein staerkeres Wachstum ermoeglicht haette. Das Defizit im Energieaussenhandel betraegt 6% vom Brutto-Inlandsprodukt gegenueber 4% in Frankreich und in der BRD. Bei gleichem Ausbau der Kernenergie wie in Frankreich haette der Energiefehlbetrag 1984 hoechstens 12 Billionen Lire (ca. 18 Mrd. DM) erreicht. Tatsaechlich betrug er infolge der Versaeumnisse 37 Billionen Lire (ca. 58 Mrd. DM). Die staatliche Elektrizitaetsgesellschaft ENEL produziert heute noch 58% des Stroms auf Erdoelbasis. In der BRD betrug der Anteil des Energietraegers Heizoel an der oeffentlichen Stromerzeugung 1984 nur noch 1%. (Atomwirtschaft/Atomtechnik 30 (1985) 8/9, S. K+U II)

BRD/Bayern:
BAYERN-ZUR HAELFTE KERNENERGIE

Anlaesslich der Mitgliederversammlung des Verbandes Bayerischer Elektrizitaetswerke wurde ein deutlich verbesserter Beitrag der Kernenergie an der Versorgung Bayerns mit Elektrizitaet und die starke Verringerung der Emissionen herausgestellt. Der Kernenergieanteil stieg von 42% im Jahre 1983 auf 49% im Jahre 1984, wobei die hervorragende Verfuegbarkeit der Kernkraftwerke entscheidend zu dieser positiven Kernenergiebilanz beigetragen hat. Eine urspruenglich fuer Mitte 1985 geplante Strompreiserhoehung durch die grossen bayrischen Elektrizitaetswerke werden unterbleiben koennen. Trotz eines Zuwachses des Stromverbrauchs um 44% seit 1976 haben sich die Emission bei Schwefeldioxid um 80% und bei Stickoxiden um 20% verringert. Bis 1989 soll sich diese Verringerung bei Schwefeldioxid auf ueber 95%, bei Stickoxiden auf ca. 75% gegenueber 1976 belaufen. (Atomwirtschaft/Atomtechnik 30 (1985) 8/9, S. K+U III)

Isotopentechnik:
RISSE IN BETON
Es wurde vom VEB Wohnungsbaukombinat Cottbus, Sitz Hoyerswerde in Zusammenarbeit mit der Technischen Hochschule Leipzig, Sektion Naturwissenschaften ein neues Verfahren zur Bestimmung von Rissen - auch visuell nicht erkennbaren, verdeckten und nur zeitweise unter Last geoeffneter Risse und von Verbundabloesungen an Industriebauten, wie Silos, Bruecken etc. sowie an Wohnungs- und Gesellschaftsbauten entwickelt. Das Wesen des Verfahrens besteht darin, auf geeignete Weise den Riss mit einer radioaktiven Fluessigkeit zu markieren. Dafuer stehen Radionuklide zur Verfuegung, die sich nur an der Rissoberflaeche anlagern, bzw. anderen mit denen auch Diffusions- und Transportvorgaenge durch poroeses Baumaterial bzw. durch Mikrorisse bearbeitet werden koennen. Der Nachweis kann mit verschiedenen Messinstrumenten erfolgen, deren Auswahl entsprechend den oertlichen Gegebenheiten optimiert wird. Die Messungen an Bauwerken nehmen ca. 5 Stunden in Anspruch. Die Durchfuehrung der Messungen erfolgt am Objekt nur von einer zugaengigen Seite. Es wird Kontaminationsfreiheit spaetestens nach 48 Stunden, in der Regel nach 12 Stunden gewaehrt. Die Einsatzmoeglichkeiten liegen in allen Bereichen, wo Risse oder Abloesungserscheinungen auftreten koennen, z.B.:
- bei der Erstellung von Bauzustandsanalysen im Wohungs-, Gesellschafts- und Industriebau
- bei der Beurteilung des Schaedigungsgrades von Bauwerken oder Bauteilen
- in der Baustofforschung
- zur Kontrolle des Haftverbundes zwischen verschiedenartigen Materialien und
- bei der Langzeitbeobachtung von Bauteilen mit geringem Schaedigungsgrad.

(Lizenzgeber: VEB Wohnungsbaukombinat Cottbus, Sitz Hoyerswerda, Kombinatsleitung, DDR-7700 Hoyerswerda, Dresdner Str. 1a).

Waldsterben:
DIE UNWISSENSCHAFTLICHE SCHAUMSCHLAEGEREI UM KKW UND WALDSTERBEN WIRD ERNEUT ANGEHEIZT
"Die radioaktiven Abgaben der schweizerischen Kernkraftwerke kommen als Ursache des Waldsterbens nicht in Frage." Diese sehr klare, wissenschaftlich fundierte Feststellung hatten die KUER (Eidg. Kommission zur Ueberwachung der Radioaktivitaet) und die HSK (Hauptabteilung fuer die Sicherheit der Kernanlagen) am 28.8.1984 aufgrund ihrer Radioaktivitaetsmessungen bekanntgegeben. Wie frueher oder spaeter zu erwarten war, haben die Atomgegner jetzt die eindeutig falsche Behauptung wieder aufgewaermt, die Kernkraftwerke koennten fuer die Waldschaeden mitverantwortlich sein. In einer Ende Juli 1985 veroeffentlichten Studie der routinemaessig fuer A-Gegner-Organisationen taetigen Zuercher Firma Infras wird der alte Unsinn wiederholt, die Streitfrage Atomkraftwerke/Waldsterben koennen heute weder mit "Ja" noch mit "Nein" beantwortet werden. Denn dazu seien zuerst weitere Forschungsarbeiten notwendig. In Anbetracht der zahlreichen in der Infras-Studie enthaltenen krassen Sachfehler ist es nicht weiter verwunderlich, dass die Autoren Ansichten kolportieren, die wissenschaftlich abgesicherten Tatsachen widersprechen. Zum Beispiel gehauptet die Infras faelschlicherweise, in mehrjaehrigen Fichtennadeln sei eine wesentlich hoehere "Anreichernung" von radioaktivem Kohlenstoff C-14 zu erwarten als die Messungen der KUER an einjaehrigen Buchenblaettern zeigten. C-14 wird aber in Pflanzen nicht - wie gewisse andere Stoffe - im Laufe der Zeit "angereichert", sondern entsprechend dem in der Umwelt vorhandenen C-14-Anteil in die Pflanzen eingebaut. Traefe die Behauptung der Infras zu, wuerde namentlich die beruehmte Methode der Altersbestimmung durch C-14-Messungen gar nicht funktionieren. Dies ist nur ein Beispiel fuer viele Luecken, mit denen die Autoren den hohen Stand ihres Nichtwissens belegen. Die Aufstellung der weiteren, teilweise ebenso krassen Fehler und Fehlinterpretationen wuerde eine lange Liste ergeben. Eigentlich muesste man fuer die Veroeffentlichung dieser politisch-ideologisch motivierten und von Sachfehlern strotzenden pseudowissenschaftlichen Bloedelei dankbar sein: Sie bestaetigt naemlich erneut, aber diesmal mit entwaffnender Offenheit, dass die Vermutung, die Kernkraftwerke seien am Waldsterben mitschuldig, wissenschaftlich absurd ist. (SVA-Bulletin (1985)14)
Nachbemerkung: Im SVA-Bulletin Nr. 14 (1985) ist ein 10seitiger Bericht von Dr. E. Stoll (Hauptabt. Sicherheit von Kernanlagen, Wuerenlingen, Schweiz) enthalten, der auf weitere Einzelheiten ueber "Die Kernkraftwerke und das Waldsterben" eingeht.

Kernbrennstoffe:
JAPAN HAELT AM ZENTRIFUGEN-ANREICHERUNGSPROGRAMM FEST
Wie der japanische Minister fuer Wissenschaft und Technologie, Reiichi Takeuchi, an einer Pressekonferenz vom 6.6.1985 im Anschluss an eine Kabinettssitzung erklaerte, wird Japan an seinem Programm zum Aufbau einer eigenen U-Anreicherungsindustrie auf der Basis von Gaszentrifugen festhalten. Die Frage einer allfälligen Ueberpruefung der japanischen Planung war nach dem Entscheid der Amerikaner vom 5.6. 1985, ihr Zentrifugen-Programm abzubrechen. Japan plant auf dem Standort Rokkashomura in der Praefektur Aomori den Bau einer kommerziellen Zentrifugen-U-Anreicherungsanlage, die vorerst eine Kapazitaet von 150 t Trennarbeitseinheiten (TEA) pro Jahr aufweisen und im Endausbau 1500 t TAE/a erreichen soll. Am 19.6.1985 erklaerte der Vorsitzende der Federation of Electric Power Companies (FEPC), die Technik der Urananreicherung mit Hilfe von Laserstrahlen koennte dereinst einer zweiten kommerziellen Anlage von 1500 t TAE/a auf dem Standort Rokkashomura zugrundegelegt werden. Wie das Japan Atomic Industrial Forum dazu feststellt, ist dies das erste Mal, dass in Japan eine fuer Planungsentscheide massgebliche Persoenlichkeit oeffentlich auf die Moeglichkeit der Kommerzialisierung der Laser-Anreicherungstechnik hingewiesen hat. (SVA-Bulletin (1985) 14)

Frankreich: U-ANREICHERUNG MIT LASERN FUER DAS NAECHSTE JAHRHUNDERT
Das franzoesische Commissariat à l'energie atomique (CEA) hat sich entschieden, die Technik der U-Anreicherung mit Lasern im Hinblick auf die Kommerzialisierung weiter zu entwickeln. Man erwartet, dass die Lasertechnik bis zur Jahrhundertwende sehr guenstige Preise fuer die U-Anreicherung ermoeglichen wird. Dies fuehrte der Praesident und Generaldirektor der Compagnie générale des matières nucléaires (Cogéma), Francois de Wissocq, am 21.6.1985 anlaesslich der Bilanz-Pressekonferenz seines Unternehmens aus. Mit der Entwicklung der Laser-Anreicherungstechnik will Frankreich, das heute mit der Anlage der Eurodif in Tricastin ueber die modernste Anreicherungskapazitaet auf Gasdiffusionsbasis verfuegt, seine Fuehrungsposition und Konkurrenzfaehigkeit auf dem Gebiet der U-Anreicherung ueber die Jahrhundertwende hinaus sicherstellen. Mit diesem Entscheiden laeuft jetzt die Entwicklung in Frankreich und in den USA parallel, waehrend Japan sein Programm zur Urananreicherung mit Gaszentrifugen

weiterfuehrt. Zustaendig fuer die Forschung u. Entwicklung auf dem Gebiet der Laser-Anreicherung, ist die Division d'etudes de separation isotopique et de chimie physique (DESICP) des CEA. Der Entscheid, einen grossen Teil der ges. Anstrengungen fuer Forschung und Entwicklung im Bereich der Isotopentrennung der Lasertechnik zu widmen, fiel bereits 1983. Wie in den USA steht in Frankreich das Verfahren der selektiven Ionisation von Atomen des Isotops U-235 in U-Metalldampf im Vordergrund des Interesses. (SVA-Bulletin (1985) 14)

Veröffentlichungen / Publications

JOINT INSTITUTE FOR NUCLER RESEARCH - DUBNA/USSR

E 4-84-446
Nuclear decay by emission of charged particle - superasymmetric fission process - 1984, 91pp., in English, D.N. Poenaru et al.

E 7-85-5
Studies of induced and natural radioactivity on heavy ion beams at Dubna - 1985, 23pp., in English

P 13-84-538
IBR-2 reactor start-up and the first neutron beam investigations - 1984, 16 pp., in Russian, V.D. Ananyev et al.

P 15-84-474
Superheavy quarks, elementary particles, nuclei - 1984, 11pp., in Russian, V.A. Karnaukhov et al.

P 10-85-333
DECAN-Program for determining parameters of exponential distributions in experiments on the investigation of genetically linked decays of radionuclides - 1985, 12pp., in Russian, H. Bruchertseifert, V.B. Zlokazov

P 6-85-211
Separation and identification of the short-lived isotopes of W from products of the reaction Gd-154+Ne-20 - 1985, 6pp., in Russian, H. Bruchertseifer et al.

UNITED KINGDOM ATOMIC ENERGY AUTHORITY

AERE-M 3412
Plotting program PLOT and subroutines for the HP7221C plotter on VAX 11/730 computer - March 1985, 67pp., J. Asher

AERE-R 11498
The investigation of Pu and Th surface contamination on steels using the Harwell nuclear microprobe - February 1985, 36pp., P.M. Pollard et al.

AERE-R 11616
A theoretical model of the swelling of gamma-irradiated epoxy resin - April 1985, 48pp., S.M. Murphy

AERE-R 11649
Pollution by wastes, edited - April 1985, 160pp., E.E. Finnecy

AERE-R 11656
Electrochemical decladding of Magnox fuel - March 1985, 34pp., T.R. Jenkins et al.

AERE-R 11736
The use of cover materials to attenuate landfill gas odour at Uglay landfill, Essex,-I. Preliminary laboratory investigation - 30pp., J.R. Emborton, P.J. Young

STAATLICHES AMT FUER ATOMSICHERHEIT UND STRAHLENSCHUTZ

SAAS-315
Zur Prophylaxe und Therapie des akuten Strahlensyndroms - 1984, 191 S., G. Bollmann et al.

SAAS-322
Die Strahlenschutzsituation bei der Verwendung von Ionisationsrauchmeldern in der DDR - 1984, 14 S., R. Ragisse

SAAS-323
Zur Sicherheitseinschaetzung von Endlagern radioaktiver Abfaelle im Untergrund unter Enwendung der Methode der Zustandsvektoren - 1985, 77 S., D. Zappe u.a.

SAAS-327
Neue Rechtsvorschriften auf dem Gebiet der Atomsicherheit und des Strahlenschutzes - 1985, 83 S.

Veranstaltungskalender – Tagungen – Lehrgänge / Calendar – Meetings – Courses

3rd IMEKO TC-8-Symposium in Theoretical Metrology, October 08-10, 1986 in Berlin GDR
Topics:
- use of fundamental physical constants and of physical effects in metrology
- metrological characteristics of optoelectronical and laser measuring devices
- metrological assurance of measuring devices and techniques in microelectronic
- computer aided calibration procedures
- measurement traceability for environmental control

Organizer:
Wissenschaftlich-Technische Gesellschaft fuer Mess- und Automatisierungstechnik der Kammer der Technik, DDR-1086 Berlin, Clara-Zetkin-Str. 115/117

Ausgewählte Beiträge / Recently published papers

ATOMNAJA ENERGIJY 59(1985)4 (in Russian)

Asymptotic Estimate of Optimal Refuelling Strategy for a Reactor - V.D. Simonov

Dimensional Stability of Structural Materials Under High Neutron Fluence Irradiation - N.K. Vasina et al.

Change of Dislocation Density and Swelling Conditions Upon Irradiating Strongly Deformed Crystals - Z.K. Saralidze

Automated Remote Control of Transplutonium Ion-Exchange Separation Processes - I.V. Tselishchev et al.

Dependence of the Average Value and Fluctuations of Absorbed Energy on Scintillator Dimensions - F.M. Zavyalkin, S.P. Osipov

Measurement of the U-236 to U-235 Fission Cross-Section in the 0.34-7.4 MeV Neutron Energy Range - B.I. Fursov et al.

Qualitative Evaluation of Field Energy of Pulsed X-Rays Backscattered in Air - V.D. Kosarev, V.P. Mukhin

Comparison of Experimental and Theoretical Values of Effective Radiation Attenuation Coefficients for Monodispersed Absorbers - V.M. Zhdanova et al.

Experimental Determination of the Proton-Induced Characteristic X-Ray Excitation Function for Thick Targets - V.F. Volkov et al.

Data Bank for Reactor Material Test Methods - N.V. Markina et al.

Analysis of the Effectiveness of Control of the Space Power Field in Reactors on the Basis of Conventional Distribution Laws - V.A. Vlasov et al

Monte-Carlo Calculations of Gamma-Ray Field Gradients - M.P. Panin

A model of crater formation under ion bombardment - V.P. Zhukov, A.V. Demidov

ATOMNAJA ENERGIJA 59(1985)5 (in Russian)
Prompt Neutrons of the Low-Energy Fission of Atomic Nuclei - B.F. Gerasimenko, V.A. Rubchenya

Measurement of Pa-231 and Am-342 Fission Cross Sections for Fast Neutrons - B.T. Fursov et al.

Measurement of Differences in Average Energy of U-233, N-235, Pu-239 and Cf-252 Neutron-Fission Spectra - V.I. Bolshov, G.N. Smirenkin

Investigation of Electrical Resistance and Micro-hardness of V and Nb for Separate and Successive Irradiation by Neutrons, -Particles and Electrons - P.K. Khabibullayev et al.

Contribution to Irradiation Creep from Dislocation Climb - V.A. Borodin, A.I. Ryazanov

A Comparative Study of Corrosion Resistance of Austenitic Steel in Li and Pb-Li Eutectic - G.M. Gryaznov et al.

Thermodynamic Interaction of UO_2 with Carbon in the Presence of Al_2O_3 SiO_2, $SiCuUC_2$ Additives - Yu.F. Khromov et al.

Radiation-Induced Decomposition of Surface-Active Materials in Water Solutions - M.V. Vladimirova et al.

Hybrid Monitoring of the Radiation Situation - A Perspective Approach for Operative Detection and Prediction of Radioactive Environmental Contamination by NPP Effluents - I.S. Eremeyev

Peculiarities in the Distribution in Ground Level Air of Radionuclides Discharged During the Regeneration of Fuel Elements - I.N. Ruzhentsova et al.

A Mathematical Model of the Temperature Field Around Wells with Radioactive Wastes and Its Verification Under Field Conditions - E.G. Drozhko et al.

Passage of Primary Protons Through Shielding Characterizing Accidental Distribution of Matter - V.G. Mitrikas et al.

On the Possibility of Determining Nuclear Fuel Characteristics by Mass Spectrometry Without Chemical Separation of Elements - V.V. Kalygin et al.

On the Effect of Neutrons on the Polarization Plane of Quartz - I.Kh. Abdukadyrova

On Three-Dimensional Calculations of a Subcritical Heterogeneous Reactor with Neutron Source - V.M. Malofeyev

Effect of Finite Dimensions of a Moderator on the Characteristics of a Slow-Neutron Pulse-Source - N.I. Alekseyev et al.

Pu Content in Soils of the Soviet Union - F.I. Pavlotskaya et al.

Control of Accident Ejections of Gaseou Radioactive Products from Atomic Power Stations by Means of TLD - I.A. Bochvar et al.

ATOM (1985) 346
Neutron beam research at Harwell - P. Schofield

Atoms for peace - G. Greenhalgh

BNFL's discharge reduction achievements - P. Mummery

Safeguards practices benefit from advances in data processing - B.H. Patrick

ATOMWIRTSCHAFT/ATOMTECHNIK 30(1985) 8/9
Inbetriebnahme des THTR-300 - E. Baust

HTR-500- Technisches und sicherheitstechnisches Konzept - J. Schoening et al.

Konzept des modularen HTR in Zweibehaelterbauweise - H. Reutler

Anwendungen und Aussichten des HTR-Moduls - M. Andler, W. Jaeger

Die zweite Reaktorgeneration in den USA - J. Bugl

Nichtverbreitung von Kernwaffen aus industrieller Sicht - W. Stoll, E. Haas

Der SNR-300 vor der Inbetriebsetzung - J. Vogel, R. Riethmueller

Superphenix kurz vor seiner Vollendung - A. Plessa et al.

Ergebnisse der Systemstudie - Andere Entsorgungstechniken - K.D. Closs, R. Papp

ATOMKERNENERGIE/KERNTECHNIK 47(1985)1
Betriebsbegleitende Ermuedungsueberwachung von Kraftwerkskomponenten - M. Miksch

Neuentwicklungen auf dem Gebiet der Ueberwachungssysteme - K. Ruthrof, V. Streicher

Elektrochemische Dekontamination von Dampferzeuger-Heizrohren - H. Operschall

Reparatur von Brennelementen fuer Leichtwasserreaktoren - H. Knaab, K. Knecht

Konditionierung von fluessigen radioaktiven Abfaellen in Zwischen- und Endlagerbehaeltern durch Vakuumtrocknung - H. Engelage et al.

Inhaerent sichere Reaktoren - Gedanken zu einem zeitgenoessischen kerntechnischen Trend - W. Seifritz

Ringkern-Stromtransformatoren fuer Wechsel- und Gleichstrom zur Messung von Strahlintensitaeten - A. Sayed, F.M. Salama

Winkel- und Energieabhaengigkeit der protoneninduzierten Roentgenausbeute an dicken Targets - H. Afridi et al.

IAEA-BULLETIN 27(1985)2
Safeguards and non-proliferation - H. Blix

IAEA technical co-operation and the NPT - C.V. Ocon

The US voluntary safeguards offer: History and scope - F.S. Houck

National system of safeguards: Experience in Japan - T. Haginoya

Close-up on safeguards training exercise in USSR

Survey of some recent non-proliferation publications - G.Delcoigne

Text of NPT

Nuclear waste disposal: Assessing near-field effects - J. Heinonen , F. Gera

Czechoslovakia: Profile of radioactive material transport - J. Štrba

Radiation protection in the workplace - F.-N. Flakus

Food irradiation: Poised for growth ... challenges ahead - L. Wedekind

Isotopenpraxis 22 (1986) 1, S. 21 bis 24

Radiation Damage and Hydrogen Bonding in Some Mono- and Disaccharides

M. M. Abou-Sekkina[1]), *M A. Sakran, A. A. Saafan*

(Chem. Dept., Faculty of Science, Tanta University, Tanta, Egypt)

Extensive measurements have been carried out on glucose and lactose molecules comprising, infrared and electronic absorption spectra both before and after exposure to an absorbed energetic γ-ray dose (5·0 × 10⁷ rd). These materials were chosen as representatives of mono- and di-saccharides. They have intramolecular hydrogen bonding of variant origin. Results obtained were in conformity, explained, assigned and discussed in details in relation with the nature and origin of hydrogen bonding in each case. Finally, a correlation between γ-radiation damage and relative strength of hydrogen bonding was put forward and established for the first time. Finally, a suggested interaction mechanism of γ-absorbed dose with the carbohydrate molecules investigated was thoroughly established.

An Glucose- und Lactose-Molekülen wurden vor und nach einer γ-Bestrahlung mit Dosen von 5·10⁷ rd IR- und Elektronenabsorptionsspektren ausgeführt. Die ausgewählten Zucker dienten als Vertreter für Mono- und Disaccharide mit intramolekularen Wasserstoffbrückenbindungen unterschiedlicher Art. Die Ergebnisse wurden entsprechend den Bandenzuordnungen in bezug auf die Natur der Wasserstoffbrückenbindungen diskutiert und zum ersten Mal eine Beziehung zwischen den γ-Bestrahlungsschäden und den relativen Bindungsstärken der Wasserstoffbrückenbindungen hergestellt. Schließlich wurde ein Wechselwirkungsmechanismus zwischen der absorbierten γ-Dosis und den untersuchten Kohlenhydratmolekülen begründet.

Keywords

chemical bonds; chemical radiation effects; disaccharides; electron spectra; gamma radiation; infrared spectra; monosaccharides; radiation chemistry; radiolysis

Introduction

Hydrogen bonding occurs between a proton donor group A – H and a proton acceptor group B, where A is an electronegative atom O, N, S, X [F, Br, I, Cl] and the acceptor group is a lone electron pair of an electronegative atom, or a π-electron orbital of a multiple bond (unsaturated) system.

Generally, a H-bond can be characterised as a proton shared by two lone electron pairs. An intramolecular H-bond occurs only when one proton donor and one proton acceptor site on the same molecule are in favorable spatial configuration.

Crystallographic studies on the structures of the constituents of carbohydrates and polysaccharides have been investigated in relative to the number of studies on the constituents of proteins and nucleic acids. Besides, they [1] deduced the following: (1) The ring-oxygen atom is invariably a H-bond acceptor. (2) The most common situation is that each hydroxyl group is associated with two H-bonds, one donor and one acceptor. (3) An environment of one H-bond, as donor only is less common. (4) An environment of three H-bonds, with one donor and two acceptors is least common. (5) In a disaccharide, there may be intromolecular H-bonding between the two residues, as found in sucrose molecule. (6) Hydroxyl groups are not involved in H-bonding, as in sucrose, whereas the H-bond length varies from 2·68 zo 3·04 Å.

For a complete understanding of this facts, more analysis need to be carried on the same degree of completion as for the sucrose structure, where neutron diffraction measurements gave the precise location of the hydrogen atoms Formula I [2]. Sucrose has two intromalecular H-bonds involving the two hydroxymethyl groups of the furanoside residue and the ring oxygen (5) and the hydroxyl O (2) of the pyranoside residue (see Formula I).

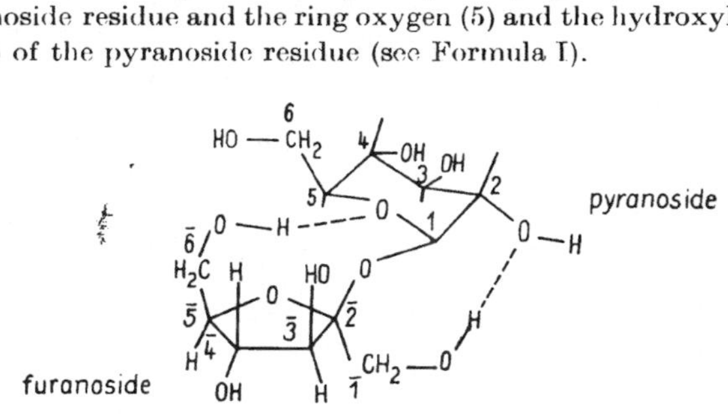

Formula I

On the other hand, the cyclic monosaccharides do not possess intramolecular H-bonding, most all of the known disaccharides display some type of intramolecular H-bonding. However, different glycosidic linkage are characterized by different intramolecular H-bonding [3].

Furthermore, while the oxygen bridge does not participate in H-bonding in any of the known glycosides. The ring oxygens of the pyranosides (in their hemiacetal form), always serve as H-bond acceptors. For the two lone electron pairs of the ring oxygen, it was found that the one directed equatorially is the one that is H-bonded [3].

Abou-Sekkina and *Assar* [4] investigated radiation damage and H-bonding in some amino-anthraquinone dyes deducing a correlation between the extent of γ-ray damage and the energy of hydrogen bonding. In another more recent publication the present authors [5] studied in details radiation damage and hydrogen bonding in biological active cyclohexane 1,2-diol and pyranol in the solid state.

[1]) Author for correspondence to be addressed

A correlation between γ-radiation damage and the relative strength of hydrogen bonding was put forward and established for the first time.

Thus, the major goal of the present manuscript is to evaluate the radiation damage of mono- and di-saccharide molecules containing intramolecular hydrogen bonding of variant origins.

Experimental

a) *Materials*
- Glucose: $C_6H_{12}O_6$, pure analytical grade supplied from El-Nasr Pharmaceutical Chemical Co., Abu-Zaabel, Egypt; anhydrous M.wt. = 180·16
- Lactose: $C_{12}H_{22}O_{11} \cdot H_2O$, pure analytical grade supplied from El-Nasr Pharmaceutical Chemical Co., Abu-Zaabel, Egypt.

b) *Infrared absorption measurements.* The IR spectra of materials investigated were carried out at room temperature using CCl_4 as a solvent (0·001 M).
A Perkin-Elmer (W. Germany) 683 infrared spectrophotometer was employed in the range 4000 – 200 cm^{-1}.

c) *Electronic absorption spectra measurements.* Solution electronic absorption spectra was followed for obtaining the room temperature electronic absorption spectra of the test samples in solution (10^{-6} mg/l in CCl_4 solution). A Unicam SP 8000 Ultraviolet Recording Spectrophotometer was utilized in the range 200 – 450 nm.

d) *γ-radiation exposure.* The materials were exposed to an absorbed gamma dose of 5·0 × 10^7 rd in the "Middle Eastern Regional Radioisotope Centre for the Arab Countries, Dokki, Cairo, Egypt" using ^{60}Co-gamma cell (activity 300 Ci) in air at room temperature.

Results and discussions

Although IR spectroscopy of pyranoid derivatives facilitated the assignments of configuration at the anomeric carbon atom in the past [6, 7], this problem is now readily tackled by NMR spectroscopy. However, it is important to realize that in contrast with the latter technique, infrared spectroscopy offers a means of obtaining data for the solid state of a particular compound, as well as for solutions.

If IR spectra are recorded at very low dilutions (<0·005 M) in CCl_4 solutions, the intramolecular hydrogen bonding associated with hydroxyl groups in carbohydrate derivatives may yield information about configuration as well as conformation. This method is based on the observation [8] that the O – H stretching frequency, which normally occurs in the region around 3630 cm^{-1}, is shifted to lower frequency when the hydroxyl group is engaged in hydrogen bonding.

The lower the frequency, the greater is the strength of the hydrogen bond. However, recent observations [9] show that care must be excercized when applying this rule.

It is well known that the ordinary D-glucose is the hemiacetal corresponding to reaction between the aldehyde group and the C-5 hydroxyl group of the open-chain structure. A powerful IR spectroscopic evidence is shown in Fig. 1, by the absence of the strong IR absorption band attributable to C=0. Although the two distinct crystalline forms of ordinary D-glucose are quite stable, in dilute solution (0·001 M) of CCl_4, each form slowly changes into the open chain structure. The process can be observed by the existance of very strong band at 1700 cm^{-1} assigned to $\nu_{C=0}$ after irradiation. This opening is of interest being as a result of the absorbed γ-dose (see Tab. 1).

In lactose molecule, the IR band at 3200 cm^{-1} before irradiation is shifted to a very strong band at 3300 cm^{-1} after irradiation which confirms strongly the formation of

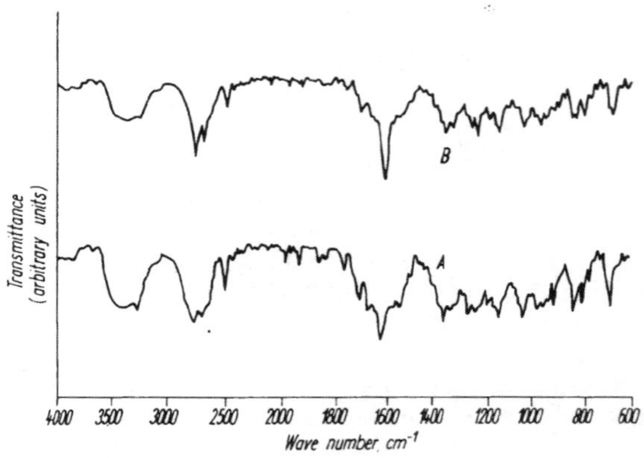

Fig. 1. The infrared absorption spectra of glucose
B – before irradiation; A – after irradiation (5 × 10^7 rd)

Tab. 1. Band assignments of the characteristic IR absorption bands of glucose and lactose molecules before (B) and after (A) the absorbed gamma dose (5·0 × 10^7 rd)

Material	m.p.	IR Absorption bands, cm^{-1}		Assignments	References
		B	A		
glucose	146 °C	3730	3730	O – H free	[10, 15]
		3460	3470	O – H free	[10, 15]
		1700	1710	$\nu_{C=0}$	[16]
		3370	3370	O – H bonded	
		2840	2840 w	C – H	[17]
		2910	2920		
lactose	200 °C	3370	3300	O – H bonded	[10, 15]
		3485	3420	O – H free	[10, 15]
		3200	2690	O – H bonded	[1, 15]
		1695	1720 w	ring carbonyl group	[10]
		1575	disappeared	$\nu_{C=0}$	
		3200	3300	O – H bonded	
			newely 1740	$\nu_{C=0}$	
		1650	1670	aldehyde, ketone	[17]
		2900	Longer wave-length		

intramolecular hydrogen bond, whereas bonded hydrogen bond absorbs at 3100–3500 cm^{-1} [10, 11]. The appearance of absorption band at 1740 cm^{-1} as assigned to new $\gamma_{C=0}$ band after irradiation confirms the oxidising action of the absorbed γ-dose (5·0 × 10^7 rd) on the secondary hydroxyl groups. Hence, this may lead to the rupture of the ring structure of lactose molecule.

A band at 1695 cm^{-1} attributable to the ring carbonyl [12] is strongly weakend by irradiation and appeared at 1720 cm^{-1}, this confirms the rupture of the ring structure. After the absorbed gamma dose, the weak band at 1650 cm^{-1} (aldehydic) [13] appeared as a strong band at 1670 cm^{-1}. This is most probably likely to be occured through the ketoenol tautomerism as previously cited [14].

In the IR spectral band corresponding to ν_s-stretching vibration of (O–H) as a result of γ-radiation damage, therefore, the hydroxyl stretching frequency in the range 3500–3000 cm^{-1} in lactose spectra is red shifted (see Tab. 1, Fig. 2).

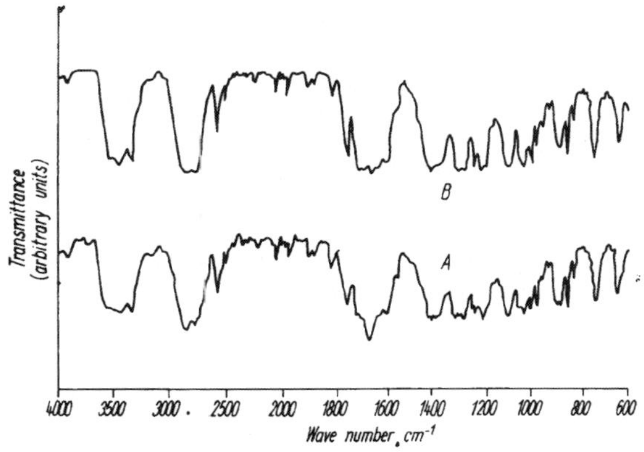

Fig. 2. The infrared absorption spectra of lactose
B – before irradiation; A – after irradiation (5·0 × 10^7 rd)

As the extent of hydrogen bonding increases the frequency of the hydroxyl band at 3370 cm^{-1} is shifted to a slightly longer wave-length of 3300 cm^{-1}. Also the band at 3485 cm^{-1} is shifted to 3420 cm^{-1} as a result of -absorbed dose (5·0 × 10^7 rd).

An examination of O–H stretching frequency around 3200 cm^{-1} showed that, a band at 3200 cm^{-1} is shifted to a very strong band observed at a longer wavelength (2690 cm^{-1}), which strengthens our suggested theory that "the formation of the intramolecular hydrogen bonding between the hydroxyl groups as a result of γ-irradiation damage of the ring structure of either glucose of galactose".

Analysis of the C–H stretching frequency regions around 2900 cm^{-1} showed similar band which shifts to longer wavelength as further support for the explanation concerning the red shift.

The ultraviolet absorption spectra ofglucose and lactose molecules was shown in Fig. 3 and Fig. 4). As a result of γ-radiation damage on electronic absorption spectra of glucose and lactose, the following could be detected in common:

1) The absorption bands are gradually shifted towards longer wavelength (red shifted). This is probably correllated with the formation of decomposition centres vibrating at lower energy lovels.
2) The intensity of some absorption bands are diminished after γ-ray absorbed dose and this is probably correlated with the γ-ray produced lattice imperfections which may lead to lattice defects and a decreased crystallinity [19].
3) Some bands disappeared and others are newely created and there is a marked broadening and weakening of the absorption bands which is probably due to the prementioned ring opening and rupture of the cyclic structure together with the prementioned decreased degree of crystallinity. All of these changes, as a result of the absorbed γ-radiation dose (5·0 × 10^7 rd), are pronouncedly occured in lactose more than in glucose molecules.

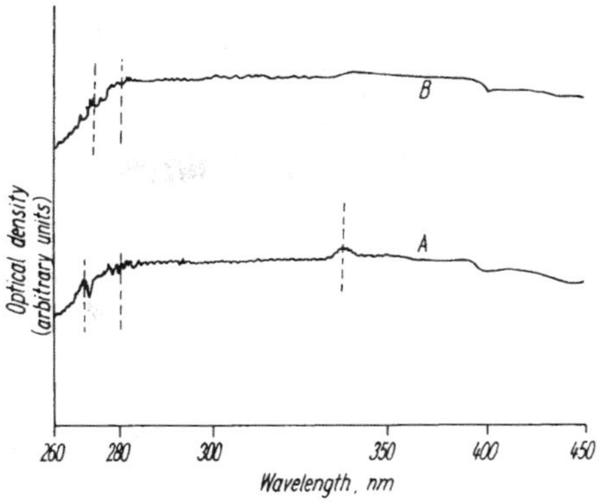

Fig. 3. The electronic absorption spectra of glucose
B – before irradiation; A – after irradiation (5·0 × 10^7 rd)

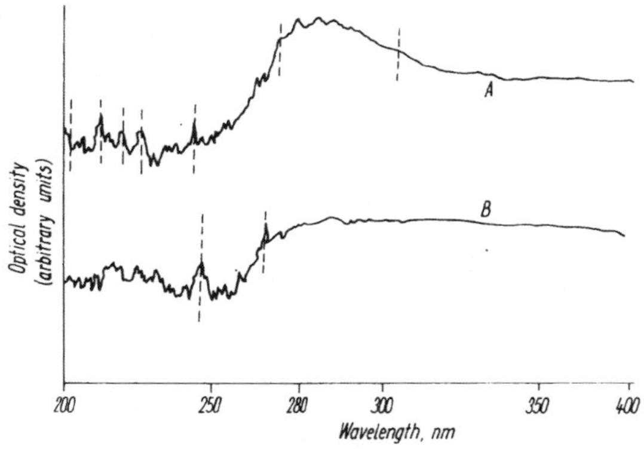

Fig. 4. The electronic absorption spectra of lactose
B – before irradiation; A – after irradiation (5·0 × 10^7 rd)

Tab. 2. The characteristic electronic absorption bands of glucose and lactose before (B) and after (A) the absorbed gamma dose (5·0 × 10^7 rd)

Material	Electronic absorption bands, nm		Assignments	References
	B	A		
		336		
glucose	272	271	Electronic transitions n – π	[16, 18]
		284		
lactose	269	278	Electronic	[16, 18]
	248	244	transitions	
		220	n – π	
	200	202		

Again these could be explained on the same prementioned attributions as being due the nature and origin of hydrogen bonding which are variant in the two investigated molecules.

Alternatively results of IR and electronic absorption spectra were almost in conformity indicating that glucose

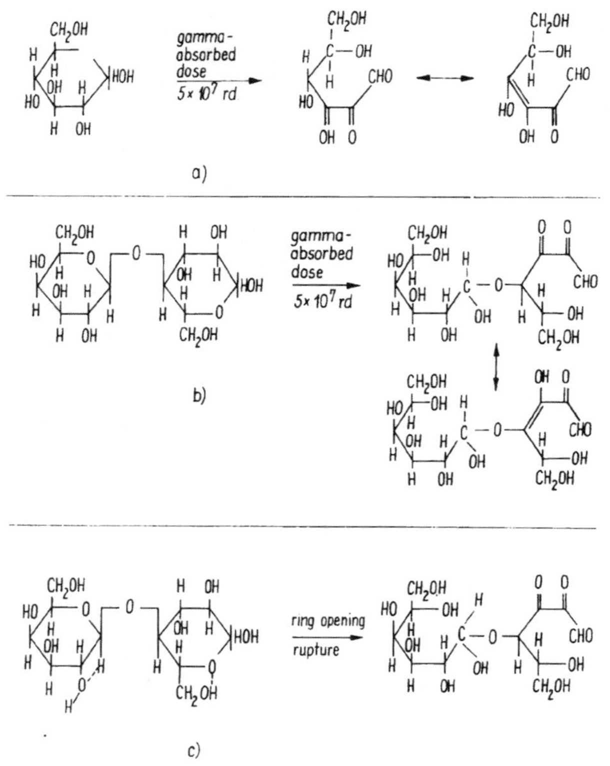

Scheme I. a) Interaction of γ-rays with glucose molecule
b) Interaction of γ-rays with lactose molecule
c) H-bonding in lactose molecule subjected to γ-absorbed dose

molecule has greater resistance towards gamma radiation damage than for lactose molecule.

Finally, a suggested interaction mechanism of gamma absorbed dose with mono-and di-saccharides was developed herein (*Abou-Sekkina*'s mechanism), and established (see Scheme I).

Received April 25, 1985

References

[1] *G. A. Jeffrey, R. D. Rosenstein,* Adv. Carbohydrate Chem. *19* (1964) 7

[2] *G. M. Brown, H. A. Levy,* Science *141* (1963) 921; *147* (1965) 1038

[3] *Andarlingam,* Biopolymers *6* (1968) 189

[4] *M. M. Abou-Sekkina, S. S. Assar,* Int. J. Appl. Rad. Isotopes *32* (1981) 847

[5] *M. M. Abou-Sekkina, M. A. Sakran, A. A. Saafan,* Isotopenpraxis *21* (1985) 281

[6] *H. Spending,* Adv. Carbohydrate Chem. *19* (1964) 23

[7] *E. L. Eliel, N. L. Allinger, S. J. Angyal, G. A. Morrison,* Conformational Analysis. New York: Wiley-Interscience 1965, p. 394

[8] *L. P. Kuhn,* J. Amer. Chem. Soc. *74* (1952) 2492

[9] *E. L. Eliel,* Accounts Chem. Res. *3* (1970) 1

[10] *R. B. Barnes, U. Liddel, V. Williams,* Anal. Chem. *15* (1943) 659

[11] *J. J. Fox, A. E. Martin,* Proc. Roy. Soc. *71* (1949) 1073

[12] *R. S. Rasmussan, R. Brittain,* J. Amer. Chem. Soc. *71* (1949) 1073

[13] *H. Hartwell, R. E. Richards, H. W. Thompson,* J. Chem. Soc. (1948) 1436

[14] *M. M. Abou-Sekkina, M. A. Sakran, A. A. Saafan,* Isotopenpraxis *22* (1986) No. 3

[15] *J. J. Fox, A. E. Martin,* Proc. Roy. Soc. (London) A *162* (1937) 419

[16] *B. J. Hendrickson, J. D. Cram, S. G. Hammond,* Organic Chemistry, 3rd Edition. New York: McGraw-Hill 1970, p. 757

[17] *J. J. Fox, A. E. Martin,* Proc. Roy. Soc. (London) A *175* (1949) 208

[18] *H. Becker, W. Berger, G. Domschke,* Organikum, Organisch-chemisches Praktikum, 15. Aufl., Berlin: VEB Deutscher Verlag der Wissenschaten 1974, 98

[19] *M. A. Moharram, S. M. Rabie,* Isotope Rad. Res. *10* (1978) 91

Isotopenpraxis 22 (1986) 1, S. 24

Buchbesprechung

Praxis nichtlinearer Gleichungen
mit zahlreichen Anwendungsbeispielen
für Ingenieure, Mathematiker und Naturwissenschaftler

W. Heitzinger, I. Troch, G. Valentin. München — Wien: Carl Hanser 1985. 364 S., 99 Abb., zahlr. Tab.; geb. DM 68,—

Wie der Titel dieses Buches bereits deutlich macht, handelt es sich um ein Buch für den Praktiker; und wer sich einmal hilfesuchend nach Lösungsmethoden nichtlinearer, insbesondere transzendenter Gleichungen umsehen mußte, wird leicht erahnen, was die Verfasser mit dieser umfangreichen Beispielsammlung im Sinn hatten, als sie dieses Buch konzipierten. Es ist in 6 übersichtlich gegliederte Kapitel geteilt, wobei zwei davon den Grundlagen (Funktionen, Nullstellen, Umformen von Gleichungen, Allgemeines über Polynome und ihre Nullstellen) gewidmet sind und die folgenden Kapitel die vielen und in der Literatur weit verstreuten Verfahren zum Inhalt haben, wie solche Gleichungen und Gleichungssysteme gelöst werden können. Der Praktiker wird zwei Vorzüge dieses Buches schätzen: die recht ausführlichen Berechnungen und die vielen aus der Praxis entnommenen Beispiele, an welchen diese Berechnungen ausgeführt werden. Daß der überwiegende Teil der Beispiele aus dem Bereich der Technik entnommen wurde, ist kein grundsätzlicher Nachteil; denn dem Untertitel nach kommt der naturwissenschaftliche Bereich etwas kurz weg, trotz der leicht auffindbaren Beispiele aus der Chemie und der chemischen Verfahrenstechnik. Wie leicht gelangt doch auch eine Folgereaktion 1. Ordnung in eine transzendente Gleichung zur Berechnung des zeitlichen Maximums aus den beiden Reaktions-Geschwindigkeitskonstanten!

Die Literatur ist mit 284 Zitaten ausführlich erfaßt, wobei das Bemühen der Autoren erkennbar ist, bis zu den Originalarbeiten vorzustoßen. So entstammt keine geringe Zahl an Zitaten dem vergangenen Jahrhundert — *A. L. Cauchy*, 1829, beispielsweise —, zurückgehend bis auf *E. Halley*, 1694. Aber rund 31% der Zitate stammen demgegenüber aus den Jahren 1975—84, erfassen also auch die moderne Literatur auf diesem Gebiet. Trotz mancher heute vorhandenen Möglichkeiten, mit einem leistungsfähigen Taschenrechner zu geeigneten Lösungen zu kommen, wird dieses Buch vielen helfen können, die sich nach eleganten numerischen Lösungsverfahren umsehen müssen. Es ist von Praktikern für Praktiker geschrieben, und so gesehen füllt es eine Lücke und wird mit Dankbarkeit zur Hand genommen werden.

H. Hübner

Isotopenpraxis 22 (1986) 1, S. 25 bis 27

Spectroscopic Studies on the Effect of γ-Irradiation on Analytical Reagents Part II — Lanthanide Acetylacetonate Complexes

A. A. Shabaka, N. Kadry[1], *R. Sharkawi*[1], *H. F. Aly*[2])

(Spectroscopy Lab. National Research Centre, [1])Faculty of Science Menia University, [2])Hot laboratory Centre, Atomic Energy Establishment, Cairo)

The effect of γ-irradiation on 9 lanthanide acetylacetonate complexes was investigated. The degradation processes as a result of γ-irradiation were followed by infrared and ultraviolet spectrometry. It was found that the rate of degradation of the different complexes in methanol solutions are always lower than in solid state. The infrared spectra of the irradiated complexes indicated that the M—O bonds are the main degraded part of the complex and the rate of decrease in the intensities of these bands depends on the type of the coordinated lanthanide. In methanol solution, the rate of degradation and the G-values with the observed doses were calculated and discussed in the light of coordination ability of the lanthanides as well as the degradation process.

Es wurden Effekte der γ-Bestrahlung von 9 Lanthaniden-acetylacetonat-Komplexen untersucht. Dabei wurden die Abbauvorgänge durch die Bestrahlung mit Hilfe der IR- und UV-Spektrometrie verfolgt. Es konnte festgestellt werden, daß die Abbaugeschwindigkeiten der verschiedenen Komplexe in methanolischer Lösung im Vergleich zur festen Phase immer langsamer sind. Die IR-Spektren der bestrahlten Komplexe zeigten, daß der Hauptanteil des Abbaus an den M—O-Bindungen erfolgt und daß die Abbaugeschwindigkeiten in den Intensitäten dieser Banden von der Art der Lanthanide abhängig sind. Für die methanolischen Lösungen wurden die Abbaugeschwindigkeiten und die G-Werte berechnet und unter dem Gesichtspunkt der Koordinationsfähigkeiten und Abbauvorgänge der einzelnen Lanthanide diskutiert.

Keywords

complexes; coordination valences; gamma radiation; g value; radiation chemistry; radiolysis; rare earths

Introduction

In hot cell chemistry, knowledge of the effect of ionizing radiation on analytical reagents is required [1]. In a programe directed to this subject, we studied the effect of γ-irradiation on some lanthanide 8-hydroxyquinoline complexes by infrared spectrophotometry [2]. In continuation of this programe, the present study is directed to investigate the effect of γ-irradiation on lanthanide acetylacetonate complexes in solid state and as solutions. Acetylacetone is well known analytical reagent which has many applications in liquid-liquid extraction as well as a complexing agent for many metal cations including lanthanides.

Experimental

The metal acetylacetonate complexes were prepared as reported previously [3]. The elemental chemical analysis was carried out by the Microanalytical unit, National Research Centre. A general chemical formula of $M(ac\ ac)_3$ for the lanthanide acetylacetonate investigated was found.

The infrared measurement were carried out using the fully automatic infrared spectrophotometer "4250 Beckman" for recording the IR spectrum. The KBr disc technique was used.

The ultraviolet absorption spectra of the investigated samples were measured as a solution in methanol. The instrument used was Beckman 5260 spectrophotometer. This instrument is a double beam grating spectrophotometer and has a resolution better than 0·05 nm.

[2]) Address: Cairo, Egypt, P.O. Box 13759

The samples were irradiated by ^{60}Co gamma 3500 NORATOM. The position of irradiation was calibrated by ferrous sulphate dosimeter.

Results and discussion

The effect of γ-irradiation were studied on the solid and liquid complexes. In the solid state the degradation of the complexes were followed by infrared spectrophotometry whereby in the liquid state, the effect of γ-irradiation were followed by UV absorption spectrophotometry.

1) *Effect of γ-irradiation on the solid complexes*

The infrared absorption spectra of the investigated lanthanide acetylacetonate complexes were measured before and after γ-irradiation with doses of 20·44, 28·62 and 36·80 Mrd. In Fig. 1, the infrared spectra of $Ho(ac\ ac)_3$ before and after γ-irradiation is given as an example.

It was found that there are no detectable change in the frequencies of the absorption bands and no new bands are observed. The main changes associated with γ-irradiation are a decrease in the intensities of certain bands in the M—O region with a relative increase in the intensities of other bands in the C—O region. The amount of decrease or increase is mainly related to the γ-dose of irradiation. Since the half band width is found to be slightly affected with increasing irradiation dose, the optical density was taken as a direct measure for the changes in intensity to measure the kinetic of the degradation of different bands.

Based on our previous work on the band assignment [4] the M—O band at 528 ± 3 and that at $405 \pm 5\ cm^{-1}$

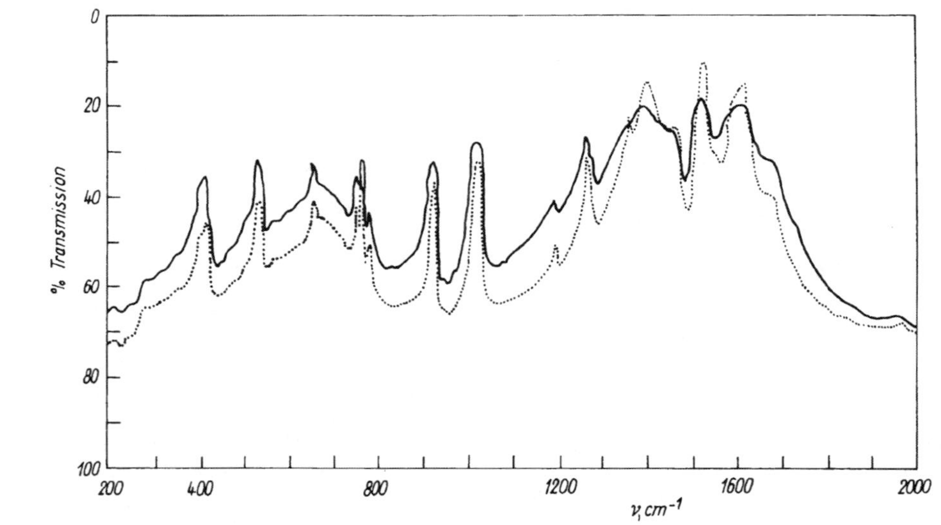

Fig. 1. IR Spectra of $Ho(ac\ ac)_3$
——— before and - - - - - - after irradiation (36·8 Mrd)

were chosen for evaluation of the γ-irradiation effects on the complexes studied, since they are directly related to the degradation of the lanthanide complexes studied.

In Fig. 2, 3 the absorbances of the two M–O bands at 528 ± 3 and at $405 \pm 5\ cm^{-1}$, respectively before and after irradiation with different doses are given as log OD — γ dose relation. In all cases, a linear decrease was obtained indicating a first order degradation process. From the slope of these lines, the rate constant of decrease, K, for the different metal complexes was calculated and given in Tab. 1.

Tab. 1 shows that the rate of decrease K [Mrd^{-1}] takes the following sequence;
For M–O at $528 \pm 3\ cm^{-1}$:

$$Tm > Sm > Gd \geqq Pr > Ce > Eu > Ho > La \geqq Er$$

Tab. 1. The rate of decrease K [Mrd^{-1}] for the M–O bands of the lanthanide (ac ac)$_3$ complexes

Complex	$K \times 10^3$ [Mrd^{-1}] for M–O band at $528 \pm 3\ cm^{-1}$	$K \times 10^3$ [Mrd^{-1}] for M–O band at $405 \pm 5\ cm^{-1}$
La(ac ac)$_3$	2·66	5·58
Ce(ac ac)$_3$	10·44	11·96
Pr(ac ac)$_3$	12·68	16·96
Sm(ac ac)$_3$	15·37	14·91
Eu(ac ac($_3$	3·52	4·96
Gd(ac ac)$_3$	12·74	10·62
Ho(ac ac)$_3$	3·31	1·72
Er(ac ac)$_3$	2·57	2·40
Tm(ac ac)$_3$	15·86	8·89

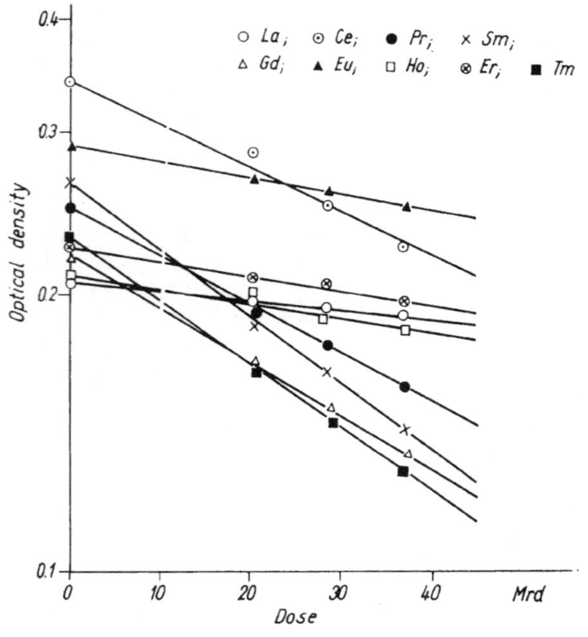

Fig. 2. Relation between optical density and absorbed dose for the M–O band at $528 \pm 3\ cm^{-1}$

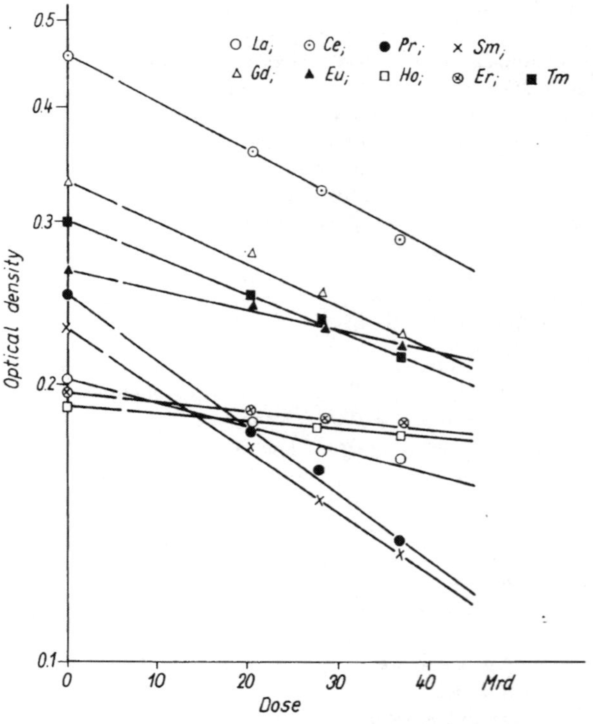

Fig. 3. Relation between optical density and absorbed dose for the M–O band at $405 \pm 5\ cm^{-1}$

For M–O at 405 ± 5 cm^{-1}

Pr > Sm > Ce > Gd > Tm > La > Eu

> Er > Ho

If the *K* values measured for the two different M–O bands are of the same origin, one should expect that the *K* values obtained from degradation of the two absorption bands for a certain complex to be the same. This was not realized except that of $Ce(ac\ ac)_3$, $Sm(ac\ ac)_3$ and $Er(sc\ sc)_3$. Therefore, it should be concluded that for the other complexes, the M–O bands are not of the same nature and one of them could be mixed with other vibronic bands.

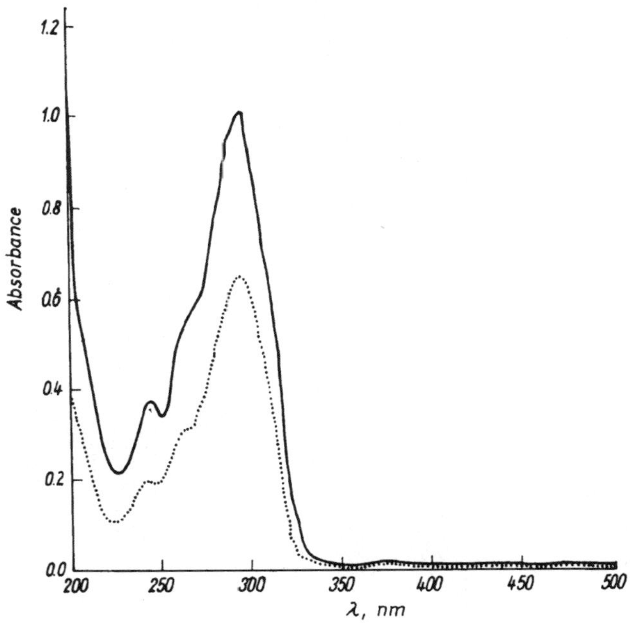

Fig. 4. UV absorption spectra of $Pr(ac\ ac)_3$
——— before and - - - - - - after irradiation (36·8 Mrd)

Tab. 2. The *K* values of the investigated complexes calculated from the ultraviolet spectra

Metal complexes	Pr	Eu	Gd	Ho	Er	Tm
$K \times 10^3$ Mrd^{-1}	1·43	5·30	2·00	3·20	2·29	6·36

2) *Effect of γ-radiation on the complexes in solutions*

The effect of γ-irradiation on liquid solution of some of the lanthanide acetylacetonate investigated was studied in methanol solution by UV spectrophotometry. Fig. 4 shows the spectrum of $Pr(ac\ ac)_3$ before and after irradiation as an example. The main band observed at (250–290 nm) was previously assigned as $\pi - \pi^*$ for the acetylacetonate molecule as a result of chelate formation [5]. Following this band as a function of absorbed dose, the intensity of this band was found to decrease by increasing the γ-dose. From the rate of decrease, the *K* values for each complex was calculated and given in Tab. 2. which takes the sequence:

Tm > Eu > Ho > Er > Gd > Pr

It is to be noted that the *K* values obtained for similar complexes in methanol solution are generally less than that in the solid state or more or less of the same order. This is quite interesting, since we should expect that the values of *K* in the solid state should be higher than that in the liquid state, since in the solid state the complexes are more condensed than in liquid solution. However, the low degradation kinetics in the liquid state could be explained in terms of the protective action of methanol on the metal complexes studied [6]. Using the molecular extinction coefficient (ε) calculated from the UV before irradiation, the concentration of the complexes at different doses was calculated using the formula:

$$C_D = \frac{(OD)_D}{\varepsilon L}$$

where C_D and *OD* are the concentration and optical density after each dose, respectively, *L* is the thickness of the cell. The *G*-value at each dose was also calculated. The results are given in Tab. 3.

This table shows that the concentration of the complex decreases by increasing the absorbed dose. On the other hand, the *G*-values obtained differ with the absorbed dose for the different complexes studied indicating that the process of degradation of the complexes differs with the absorbed doses.

Receved February 6, 1985
Accepted in reviesd form April 26, 1985

References

[1] *D. C. Stewart, H. A. Elion,* (Eds.), Progress in Nuclear Energy IX, Analytical chemistry Vol. 10, chapter 14, on "Effect of Radiation on Analytical Reagents" by *H. F. Zittel*, Pergamon press 1970

[2] *H. F. Aly, F. M. Abdel-Kerim, A. El-Agramy, M. H. Atya,* Isotopenpraxis *13* (1977) 59

[3] *D. Prushotham, V. R. Rao, S. V. R. Raos,* Anal. Chem. Acta *33* (1965) 162

[4] *H. F. Aly, F. M. Abdel-Kerim, M. M. El-Dessoky* and *A. M. El-Agramy,* Radiochem. Radio anal. Letters *28* (1977) 109

[5] *D. W. Barnum.* J. Inorg. Nucl. Chem. *12* (1961) 221

[6] *J. H. O'Donnel, D. F. Sangster* "Principles of Radiation Chemistry" London: Edward Arnold Publisher 1970

Tab. 3. The concentration and *G*-values of the investigated complexes at different doses

Metal Complexes	$C \times 10^{-4}$ mol/l	$C \times 10^{-4}$ mol/l 20·44	*G*	$C \times 10^{-4}$ mol/l 28·62	*G*	$C \times 10^{-4}$ mol/l 36·80	*G*	$\varepsilon \times 10^{-4}$ $mol^{-1} \cdot l \cdot cm^{-1}$
$Pr(ac\ ac)_3$	0·60	0·57	0·15	0·55	0·17	0·39	0·57	1·67
$Eu(ac\ ac)_3$	0·53	0·52	0·05	0·46	0·25	0·45	0·22	2·79
$Gd(ac\ ac)_3$	0·26	0·24	0·01	0·23	0·11	0·21	0·14	4·00
$Ho(ac\ ac)_3$	0·26	0·24	0·01	0·21	0·17	0·20	0·16	2·23
$Er(ac\ ac)_3$	0·52	0·50	0·01	0·46	0·21	0·41	0·30	2·88
$Tm(ac\ ac)_3$	0·72	0·57	0·73	0·55	0·59	0·26	1·25	1·13

Isotopenpraxis 22 (1986) 1, S. 28 bis 31

Neue Sonden für das Flächenmassemeßsystem 24024*)

R. Bernhardt, W. Dietzsch

(VEB Robotron-Meßelektronik „Otto Schön" Dresden)¹)

Für das radiometrische Flächenmassemeßsystem 24024 des VEB Robotron-Meßelektronik „Otto Schön" Dresden wurden neue Meßwertaufnehmer entwickelt. Das neue Sortiment von Durchstrahlsonden umfaßt 21 für die Lösung unterschiedlicher Meßaufgaben geeignete Varianten. Gegenüber den Vorläufersonden wurden die Meßsignalstabilität und das Zeitverhalten verbessert, der Gesamtmeßbereich erweitert und die Auflösung bei Profilmessungen erhöht. Darüber hinaus liegen verbesserte Strahlenschutzkennwerte vor.

To the radiometric thickness gage robotron 24024 manufactured by the VEB Robotron-Meßelektronik „Otto Schön" Dresden new sensors basing on absorption of beta and gamma radiation were developed. Altogether 21 modifications of one base type suitable for the solution of different measuring problems are available. In comparison to previous sensors the stability of the measuring signal has been advanced. Even, there are a diminished response time and enlarged thickness respectively mass per unit area ranges. Profile measurements with a higher resolution can be carried out. Moreover, the radiation protection was improved.

Keywords

industry; isotope applications; radiometric gages; thickness

1. Einführung

Im VEB Robotron-Meßelektronik „Otto Schön" Dresden steht die Entwicklung einer neuen Generation von radiometrischen Meßwertaufnehmern für die Flächenmasse bzw. Materialdicke vor dem Abschluß. Diese neue Sondengeneration löst bis auf die Rückstreu- und Dichtemeßsonden die bisher im Flächenmassemeßsystem 24024 sowie als Bestandteil der Flächengewichtsmeßanlage 24004 verwendeten Meßwertaufnehmer ab. Über radiometrische Meß- und Regeleinrichtungen des VEB Robotron-Meßelektronik „Otto Schön" Dresden, die in verschiedenen Industriezweigen bei der Herstellung band- und tafelförmiger Halbzeuge sowie Fertigprodukte zum Einsatz kommen, wurde an dieser Stelle schon mehrfach berichtet [1–4].

Eine radiometrische Meßanlage für die berührungslose Flächenmasse- bzw. Dickenbestimmung bei laufender Fertigung setzt sich im allgemeinen aus den Hauptbaugruppen

- Sonde (mit Haltevorrichtung)
- Bewegungseinrichtung sowie der
- Auswerte- und Bedieneinheit

zusammen. Abb. 1 zeigt eine typische Baugruppenkonfiguration für eine Durchstrahlmeßanlage. Die Sonde besteht aus einem Strahlkopf als Sender und einem Meßkopf als Empfänger. Beide Funktionseinheiten sind an einem Bügel befestigt, in den das Meßgut eintaucht. Den periodisch erfolgenden Transport der Sonden von der Meßposition in die Abgleichstellung (Nutzstrahlbündel außerhalb der Meßgutbahn) übernimmt eine Ausfahreinrichtung, auf der der Meßbügel fahrbar gelagert ist.

In einer radiometrischen Meßeinrichtung fällt der Sonde (zusammen mit der Halterung und Bewegungseinrichtung)

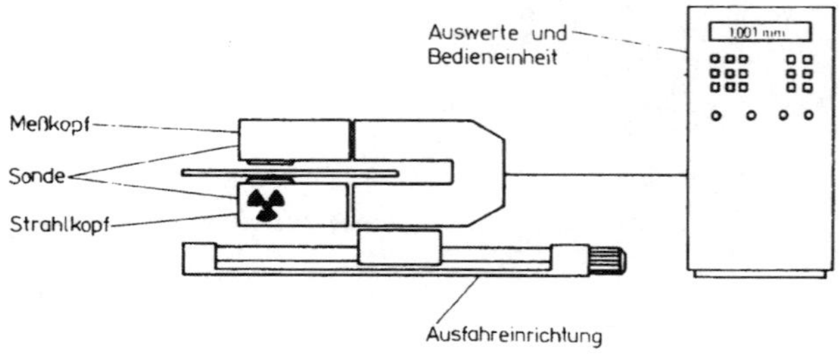

Abb. 1. Typische Baugruppenkonfiguration für eine radiometrische Durchstrahlmeßanlage

die Aufgabe zu, ein elektrisches Meßsignal bereitzustellen, aus dem die Auswerteeinrichtung mit Hilfe des gespeicherten funktionalen Zusammenhangs (Kalibrierfunktion) den Flächenmasse- bzw. Dickenmeßwert oder Abweichungen dieser Meßgrößen von einem vorgegebenen Sollwert ermittelt.

2. Allgemeine Entwicklungsforderungen

Wichtige Leistungsparameter, deren Verbesserung bei einer Sondenentwicklung im Mittelpunkt steht, sind die Stabilität und das Zeitverhalten des Meßsignals, der Meßbereich, die Größe und Geometrie des Meßflecks sowie die Strahlenschutzkennwerte.

Kennzeichnend für die Stabilität des Meßsignals sind die Beträge der Abweichungen vom Kalibrierwert, d. h. von dem Signal, das sich gemäß gespeicherter Kalibrierfunktion für die vorliegende Flächenmasse ergeben müßte. Ursache für solche Abweichungen können einerseits statistische Schwankungen, andererseits sich längerfristig auswirkende Driften von Bauelementeparametern sein (Aktivitätsänderung von Strahlungsquellen, Drift des Arbeitswiderstandes u. a. m.). Um den Einfluß von Driften zu minimieren, wählt man als Meßsignal M einen auf das Abgleichsignal (Volleinstrahlungssignal) bezogenen

*) Vortrag, gehalten auf dem 7. Kolloquium Technische Anwendung radioaktiver Isotope im Berg- und Hüttenwesen, 13.–14. 3. 85, Freiberg

¹) Anschrift: DDR-8012 Dresden, PSF 211

Relativwert. Ausschlaggebend für die erforderliche Länge einer Abgleichperiode ist die Langzeitstabilität des Meßsignals. Eine Abweichung ΔM überträgt sich auf den relativen Fehler ($\Delta s/s$) des ermittelten Flächenmassewertes s entsprechend der physikalischen Empfindlichkeit $[\mathrm{d}M/(\mathrm{d}s/s)] = -\mu s \exp - \mu s$. Diese Beziehung erhält man durch Differentiation des exponentiellen Schwächungsgesetzes mit dem Massenschwächungskoeffizienten μ, das beim Durchstrahlverfahren näherungsweise für die Kalibrierfunktion gilt.

Aus der Beziehung für die physikalische Empfindlichkeit geht hervor, daß der Meßbereich (für ein bestimmtes Meßgut) von Art und Energie der eingesetzten Strahlung abhängt. Bei gegebener Strahlungsart verschieben sich die Meßbereichsgrenzen, festgelegt durch einen vorgegebenen Minimalwert für die physikalische Empfindlichkeit, mit wachsender Strahlungsenergie zu größeren Flächenmassen. Insbesondere aus der Sicht des Geräteherstellers ergibt sich daraus als konstruktive Forderung die Austauschbarkeit der Strahlungsquelle, so daß die Sonde zur Abdeckung eines möglichst großen Gesamtmeßbereichs einsetzbar ist.

Als Meßfleck bezeichnet man den von Strahlung durchsetzten Flächenbereich des Meßgutes. Im Hinblick auf die Anwendung der Sonden bei schmalen Bändern und auf eine hohe Auflösung bei Profilmessungen sind möglichst kleine Meßfleckdurchmesser anzustreben. Andererseits ist bei bestimmten Meßgütern mit inhomogener Flächenmasseverteilung (z. B. bestimmten textilen Materialien) die integrierende Wirkung bei Durchstrahlung größerer Flächenbereiche erwünscht. Unabhängig davon sind der Meßfleckverkleinerung Grenzen gesetzt, weil mit engerer Kollimierung des Nutzstrahlbündels, d. h. mit abnehmendem Öffnungswinkel und damit abnehmender Zahl der das Meßsignal erzeugenden Strahlungspartikel, die Größe der statistischen Meßsignalschwankungen zunimmt und dem nicht beliebig durch Erhöhung der Quellenaktivität oder Detektorempfindlichkeit engegengewirkt werden kann. Bei Querprofilmessungen wird daher zunehmend mit schlitzförmigen Kollimatoren (und dem angepaßter Quellen- und Detektorgeometrie) gearbeitet. Durch den Übergang zur Rechteckgeometrie beim Meßfleck läßt sich ohne Verringerung des durchstrahlten Raumwinkels eine höhere Auflösung in Richtung der Schmalseite des Meßflecks erzielen.

Hauptzielstellung im Strahlenschutz ist es, die Strahlenemission der Sonden möglichst so weit herabzusetzen, daß bei Installation der Meßanlage an einer Fertigungseinrichtung keine ständigen Arbeitsplätze des dafür und für die Meßeinrichtung zuständigen Bedienpersonals in den Kontrollbereich fallen, daß also keine dosimetrischen Überwachungsmaßnahmen erforderlich werden. Darüber hinaus ist die Konstruktion des Strahlkopfes auf die Minimierung der Strahlenbelastung bei Reparatur- und Wartungsarbeiten, insbesondere bei der Quellenmontage, auszurichten.

3. Eigenschaften der neuen Sonden M 7200

3.1. Sortimentsübersicht

Das neuentwickelte Sortiment von Durchstrahlsonden umfaßt 21 auf die Lösung unterschiedlicher Meßaufgaben zugeschnittene Varianten (zu den Hauptkennwerten s. Tab. 1).

Im Grundaufbau stimmen sämtliche Sondenvarianten überein. Strahl- und Meßköpfe besitzen Gehäuse aus Aluminiumguß mit einheitlichen, von den Vorläufersonden übernommenen Außenabmessungen (470 mm × 216 mm × 170 mm). Dem Einsatz in Walzwerken entsprechend sind die Sondengehäuse öldicht ausgelegt (Schutzgrad IP 67 gem. TGL RGW 778). Die Strahleneintritts- bzw. Strahlenaustrittsfenster von Meß- und Strahlkopf werden durch ein Abweiserblech aus Stahl vor Zerstörung durch berührendes Meßgut (z. B. beim Durchlauf eines Bandendes durch den Meßspalt) geschützt. Wie am Beispiel des in Abb. 2 dargestellten Strahlkopfes ersichtlich ist, bedeckt der Abweiser die gesamte dem Meßgut zugewandte Gehäuseseite und schützt damit auch die darunter montierten Abblasdüsen für die Beseitigung störender Beläge (z. B. Walzöl) auf den Fenstern mittels Preßluftstrahl. Im Rahmen der Weiterentwicklung der Sonden wurde für die

Tab. 1. Technische Daten der neuen Sonden M 7200

Typ-Nr.	Strahlungsart	Nuklid	Aktivität GBq	Bügelweite mm	Meßfleckgeometrie	Zeitkonstante, ms	Meßbereich gm^{-2}	Explosionsschutz
M 7200·10			18,5	40	rund	20		
M 7200·11		^{85}Kr	8,5	40	rechteck.	50	45 bis 1250	
M 7200·12			5,5	40	rund	50	(bez. auf Al)	(Ex)fü u. (Ex)f
M 7200·13			8,5	40	rechteck.	50		(Ex)fü u. (Ex)f
M 7200·14	Betastrahlung		3,7	60	rund	20		
M 7200·15			3,7	40	rund	20		
M 7200·16			3,7	100	rund	20		
M 7200·17		^{90}Sr-^{90}Y	2 × 1,1	40 u. 60	rechteck.	50	180 bis 6400	
M 7200·18			1,1	60	rund	50	(bez. auf Al)	(Ex)fü u. (Ex)f
M 7200·19			0,9	40	rund	50		(Ex)fü u. (Ex)f
M 7200·20			1,8	100	rund	50		(Ex)fü u. (Ex)f
M 7200·21			2 × 1,1	40 u. 60	rechteck.	50		(Ex)fü u. (Ex)f
M 7200·22	Bremsstrahlung	^{85}Kr	130	60	rund	50		
M 7200·23			150	100	rund	50	1330 bis 67 500	
M 7200·24			130	60	rund	50	(bez. auf Fe)	(Ex)fü u. (Ex)f
M 7200·25			150	100	rund	50		(Ex)fü u. (Ex)f
M 7200·26			10 bis 100	60, 100 u. 150	rund u. rechteck.	20 u. 50		
M 7200·27	Gammastrahlung	^{241}Am	10 bis 100	60, 100 u. 150	rund u. rechteck.	20 u. 50	445 bis 47 000 (bez. auf Fe)	(Ex)fü u. (Ex)f
M 7200·28			5 × 20	150	rund	20		
M 7200·29			14	150	rund	50		
M 7200·30			2 × 14	60	rechteck.	50		

Abb. 2. Strahlkopf mit Abweiser

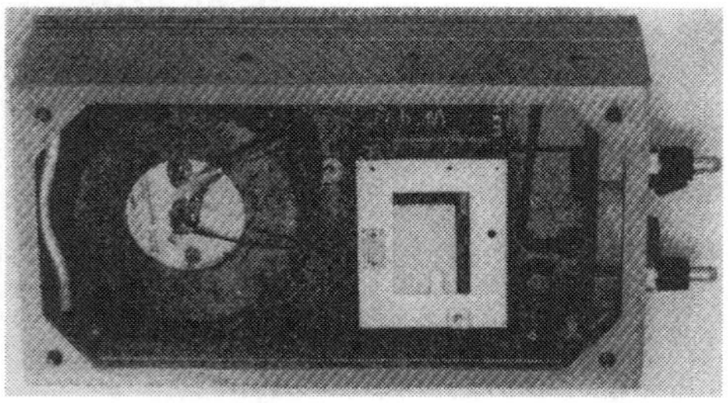

Abb. 3. Innenansicht des Meßkopfes mit Gleichspannungsverstärker und abgeschirmter Ionisationskammer

Abb. 4. Innenansicht des Strahlkopfes mit dem Arbeitsbehälter

Abblasdüsen bei gleichzeitig verbesserter Wirksamkeit eine fertigungsgerechtere Lösung erarbeitet. Zu den allen Varianten gemeinsamen Funktionsgruppen gehören der Gleichspannungsverstärker im Meßkopf und der die Strahlungsquelle aufnehmende Arbeitsbehälter mit Verschlußmechanismus im Strahlkopf (s. Abb. 3 und 4). Der im Meßkopf montierte Verstärker mit dem thermostatisierten Arbeitswiderstand im Gegenkopplungszweig gibt als analoges Meßsignal Spannungswerte zwischen 0 und 20 V ab. Der Arbeitsbehälter aus Hartblei enthält eine elektromagnetisch angetriebene Walze mit einem Quellenschacht, dessen Ausgang sich bei freigegebenem Strahlenbündel vor der Strahlenaustrittsöffnung in der äußeren Ummantelung des Arbeitsbehälters befindet. Der die Walze bewegende Drehmagnet arbeitet im Betriebsfall ständig gegen eine Rückstellfeder, so daß bei Havarien mit Ausfall der Versorgungsspannung eine automatische Abblendung des Strahlenbündels gesichert ist.

Unterschiede zwischen den Sondenvarianten bestehen in bezug auf die Art der Strahlung, die Aktivität der Strahlungsquelle, den Wert des Arbeitswiderstandes, das Material der Strahlenein- und -austrittsfenster, die Ausführung der als Strahlungsdetektoren eingesetzten Ionisationskammern sowie die Dimensionierung der Kollimatoren vor der Austrittsöffnung am Arbeitsbehälter. Ein Teil der Sondenvarianten ist mit Belüftungsschlangen ausgerüstet, die Explosionsschutz entsprechend den Schutzarten Fremdluftüberdruck und Fremdbelüftung ((Ex) fü T4 und (Ex) f T4 nach TGL 19491) gewährleisten. Die Strahlenein- und -austrittsfenster bestehen aus Titan- oder Aluminiumfolie, je nachdem ob Beta- oder Photonenstrahlung eingesetzt wird. Diesen Strahlungsarten angepaßt werden auch unterschiedliche Ionisationskammern verwendet. Die Dimensionierung der Kollimatoren richtet sich nach der Meßspalt- bzw. Bügelweite und der Meßfleckgeometrie.

Gegenüber den Vorläufersonden wurden der Meßfleckdurchmesser von ca. 150 mm auf etwa 50 mm verringert und Varianten mit Rechteckgeometrie des Meßflecks (ca. 20 mm × 50 mm) eingeführt.

Die Weiterentwicklung der Sonden erstreckte sich zum Teil auch auf das Zubehör. Beim Bügelsortiment erfolgte eine Aufstockung um Bügel mit den Weiten 100 mm und 150 mm (bisherige Bügelweiten: 20 mm, 40 mm und 60 mm). Durch neu geschaffene Verlängerungsstücke ergeben sich zusammen mit den in zwei Längen vorliegenden Bügeln zusätzliche Möglichkeiten, die Eintauchtiefe des Meßgutes in den Meßspalt zu variieren (mögliche Eintauchtiefen: 425 mm, 595 mm, 850 mm und 1020 mm). Dem Ziel, die Reproduzierbarkeit der Meßgeometrie zu erhöhen, diente die Einführung von Paßstiften zur Fixierung der am Bügel montierten Meß- und Strahlköpfe. Abb. 5 zeigt einen komplett montierten Meßbügel mit Verlängerungselementen. Für den Transport und die Aufbewahrung der Strahlungsquellen wurde ein für alle Quellentypen und -aktivitäten einheitlicher Behälter geschaffen.

Abb. 5. Komplett montierter Meßbügel mit neuen Sonden und Verlängerungselementen

3.2. *Meßgenauigkeit und Zeitverhalten*

Im Rahmen einer weiteren Erhöhung der Signalstabilität und damit der Meßgenauigkeit wurden bei einigen Varianten des Sondensortimentes die statistischen Schwankungen des Ausgangssignals (in Relation zur Zeitkonstanten) verringert und Maßnahmen zur Kompensation des Fehlereinflusses von Änderungen der Umgebungstemperatur getroffen.

Zur Verringerung der statistischen Schwankungen wurden bei den betreffenden Sondenvarianten so weit erhöhte Aktivitäten eingesetzt, daß bei einem von 5 GΩ auf 1 GΩ verringerten Wert des Arbeitswiderstandes bei voller Einstrahlung bzw. einem bestimmten Flächenmassewert etwa die vorherigen Signalspannungswerte erreicht werden. Mit der Verringerung des Arbeitswiderstandes hat sich die für das Ausgangssignal der Sonden geltende Zeitkonstante τ, die für die gesamte Flächenmassemeßanlage einen nicht unterschreitbaren Wert darstellt, von früher ca. 50 ms auf einen Wert kleiner 20 ms verringert. Mit dieser Verbesserung des Zeitverhaltens der Sonden wird dem Trend zu höheren Fertigungsgeschwindigkeiten Rechnung getragen, der ein schnelleres Ansprechen zur Prozeßregelung eingesetzter Meßanlagen auf Änderungen der Flächenmasse bzw. Materialdicke erfordert.

Zur Kompensation des Fehlereinflusses von Änderungen der Umgebungstemperatur wurden zweierlei Maßnahmen getroffen. Zum einen erfolgte einheitlich für alle Sondenvarianten der Einbau von Kühltaschen in die Seitenwände der Gehäuse von Meß- und Strahlkopf. Gegenüber den früher an den Gehäuseinnenwänden entlanggeführten Kühlschlangen aus Kupferrohr hat sich der Wärmeaustausch zwischen dem Kühlwasser und den temperaturempfindlichen Funktionsgruppen so verbessert, daß über die Kühlung hinaus eine wirksame Thermostatisierung der Sonden möglich geworden ist. Bei Konstanthaltung der Sondentemperatur mit Hilfe des Digitalthermostaten U 16 D (Hersteller: VEB MLW Prüfgerätewerk Medingen) wurden bei einer Temperatur von 45 °C im geschlossenen Wasserkreislauf Regelfaktoren >10 erreicht, d. h., die durch Änderung der Umgebungstemperatur bewirkten Temperaturänderungen an kritischen Stellen in der Sonde (an der Ionisationskammer, in den eingeschlossenen Luftstrecken) betragen weniger als 10% der Umgebungstemperaturänderung.

Durch Regelung der Sondentemperatur allein lassen sich temperaturbedingte Meßfehler nicht vollständig kompensieren, weil dadurch die Temperaturänderungen im Meßspalt, die zu Änderungen der Luftdichte führen, nicht erfaßt werden. Zur Korrektion des Fehlereinflusses temperaturbedingter Dichteschwankungen der Luft im Meßspalt wurde daher ein Meßfühler entwickelt, der ein temperatur- und damit luftdichteabhängiges Meßsignal liefert, das durch die Auswerteeinheit des Flächenmassemeßsystems 24024 zur rechnerischen Korrektur des Flächenmassemeßwertes herangezogen wird.

Bei gleichzeitiger Anwendung beider Kompensationsverfahren wird eine Reduzierung des Temperaturfehlers des ermittelten Flächenmassewertes auf weniger als 20% erreicht. Die Zeitkonstante des Temperaturmeßfühlers liegt in der Größenordnung von Minuten, so daß längerfristige, tageszeitlich bedingte Umgebungstemperaturänderungen vor Ort erfaßt werden können.

3.3. *Strahlenschutz*

Durch konstruktive Maßnahmen und zusätzliche am Arbeitsbehälter sowie über der Ionisationskammer angebrachte Abschirmungen konnten für die Ortsdosisleistung in 0,1 m Abstand von der Sondenoberfläche bei abgeblendeter Strahlungsquelle Werte kleiner 10 µGy/h erzielt werden. Die Ionisationskammer wird durch eine Glocke aus Hartblei abgedeckt (s. Abb. 3). Am Gehäusedeckel über der Ionisationskammer ist eine weitere Bleiabschirmung befestigt, um die senkrecht durch die Ionisationskammer und die Öffnung der Bleiglocke hindurchtretende Strahlung abzuschwächen.

Durch die Abschirmmaßnahmen konnten trotz Aktivitätserhöhung bei einigen Sondenvarianten die Kontrollbereiche gegenüber den Vorläufersonden im Durchschnitt etwa auf 25% reduziert werden.

Die sich bei maximal zulässigen Dosisleistungen von 7,5 µGy/h für Sonden mit Beta-Strahlungsquellen und 2,5 µGy/h für Sonden mit Photonenstrahlungsquellen ergebenden Abstände vom Strahlkopf bewegen sich zwischen 0,5 m (M 7200·26 bis M 7200·30) und 1,8 m (M 7200·14 bis M 7200·21). Die Werte gelten für den Betriebszustand bei aufgeblendetem Strahlenbündel und im Meßspalt vorliegendem Meßgut. In Anbetracht der geringen Ausdehnung dürfte es bei Anwendung der neuen Sonden in den meisten Fällen vermeidbar sein, daß ständige Arbeitsplätze in den Kontrollbereich fallen.

Die Strahlungsquellen werden bei den neuen Sonden in eine Fassung aus nichtrostendem Stahl montiert. Diese als Adapter dienende Fassung besitzt ein Außengewinde und wird mit Hilfe eines Magnetschlüssels in den Quellenschacht der drehbaren Walze des Arbeitsbehälters geschraubt. Der zu den Sonden mitgelieferte Magnetschlüssel schirmt den Benutzer wirksam gegen die zu montierende Strahlungsquelle ab.

Bei den Sondenvarianten für Betastrahlung befinden sich die Strahlungsquellen nicht mehr wie früher unmittelbar hinter dem Kollimator nahe am Strahlenaustrittsfenster, sondern sind ca. 40 mm tiefer im Quellenschacht der drehbaren Walze des Arbeitsbehälters eingebaut. Dadurch ist eine erhöhte Havariesicherheit (z. B. bei Riß des Meßgutbandes) gewährleistet.

Als Zubehör erfüllt der Transportbehälter ebenfalls die Bedingung, daß die Ortsdosisleistung in 0,1 m Abstand von der Behälteroberfläche maximal 10 µGy/h beträgt. Für den Quellenversand im Transportbehälter steht eine gesonderte Verpackung zur Verfügung, in der der Behälter dem Typ A entsprechend der Anordnung über den Transport radioaktiver Stoffe genügt. Ebenso kann der Quellenversand im Strahlkopf mit einer Verpackung gemäß Typ A erfolgen. In diesem Fall wird durch Verschluß des Kollimators durch einen Bleizylinder der Strahlkopf gegen Austritt von Strahlung zusätzlich abgesichert.

4. Schlußbemerkungen

Im Interesse einer ökonomischen Fertigung verfolgt der Gerätehersteller das Ziel, möglichst große Stückzahlen bei geringer Typenvielfalt abzusetzen. Große Stückzahlen können aus vielseitiger Anwendbarkeit resultieren. Als Konstruktionsprinzip für die neuen Sonden M 7200 wurde ein gemeinsamer Grundaufbau gewählt, der sich durch austauschbare Konstruktionselemente den unterschiedlichen Meßaufgaben anpassen läßt. Dadurch ist für nahezu jeden Anwender eine optimal preisgünstige Lösung zu finden.

Mit der Verfügbarkeit der neuen Sonden M 7200 lassen sich die vielfältigen Gerätefunktionen der mit dem Mikrorechner K 1520 ausgerüsteten Auswerte- und Bedieneinheit des Flächenmassemeßsystems 24024 noch wirkungsvoller zur Geltung bringen.

Eingegangen am 3. 4. 1985

Literatur

[1] *R. Kießling*, Isotopenpraxis *13* (1977) 248
[2] *W. Dietzsch*, Isotopenpraxis *17* (1981) 349
[3] *W. Dietzsch, K.-H. Reinhardt*, Isotopenpraxis *19* (1983) 11
[4] *R. Bernhardt*, Isotopenpraxis *20* (1984) 4

Isotopenpraxis 22 (1986) 1, S. 32 bis 35

Automatisierung der Meßwerterfassung und -verarbeitung eines NMR-Spektrometers und Integration in ein Lokales Computernetz

B. Lindemann, H. Sprinz, U. Hädrich, R. Jung, G. Sack

(Zentralinstitut für Isotopen- und Strahlenforschung der AdW)[1])

Zur Steuerung des Meßdurchlaufes sowie der Übernahme und Vorverarbeitung der Meßdaten wurde das NMR-Spektrometer BS 487 C/TESLA mit einem Mikrocomputer AMCA-80 on-line gekoppelt. Dieses Meßsystem wurde in das bereits vorhandene Lokale Computernetz ATLAS des ZfI integriert. Zur Kopplung und Datenübertragung wurde das System CAMAC eingesetzt. Umfangreiche zeit- und speicherintensive Auswerteprogramme werden an der EDVA EC 1040 abgearbeitet. Die Meßdatenerfassung und -verarbeitung erfolgt im Dialog unter Realzeitbedingungen.

For the control of measurements as well as for taking over and pre-processing of the measurement data the NMR-spectrometer BS 487 C/TESLA has been connected on-line with an AMCA 80-microcomputer. This experimental equipment has been integrated in the local computer network ATLAS. For the data transmission and the computer coupling the CAMAC-system has been used. The storage and time intensive programmes for the spectra evaluations are running in the EC-1040 computer. The data transfer is controlled by the help of a two-party, cooperative communication type protocol. The data acquisition and processing work in dialogue are running under real-time conditions.

Keywords

data transmission; distributed data processing; nmr spectrometers; on-line systems; real time systems

1. Einführung

Das NMR-Spektrometer BS 487 C/TESLA wird für die Isotopen- und Reinheitsanalytik der im ZfI produzierten deuterierten Verbindungen eingesetzt. Dafür ist eine exakte, dem internationalen Stand entsprechende, Analytik erforderlich, um auch schwache Verunreinigungen, die sich bei der Anwendung störend bemerkbar machen, nachweisen zu können.

Dies erfordert die Verbesserung des Signal-Rausch-Verhältnisses durch eine Spektrenakkumulation mit Nullinienkorrektur und der Vermeidung der Drift wichtiger Geräteparameter. Die Auswertung der Meßdaten erfolgt mit Methoden der Isotopenanalyse. Für ein strahlenbiologisches Forschungsprogramm [2] sind außerdem umfangreiche Linienformanalysen am Großrechner notwendig, die im off-line-Betrieb einen sehr großen Zeitaufwand benötigen.

Für diese Anwendungsgebiete ergeben sich durch die Realisierung einer on-line-Datenauswertung Möglichkeiten zur Verbesserung der Signalanalyse und der Rationalisierung der Forschungsarbeit. Der genannte Spektrometertyp ist zwar nicht für eine Rechnerkopplung konzipiert, aber durch die Nutzung der Schreiberfunktionen kann eine rechnerseitige Steuerung der Spektrenaufnahme mittels eines speziellen Adapters realisiert werden. Da das Spektrometer über NF-Generatoren hoher Konstanz verfügt ($\pm 0{,}2$ Hz in 3 Stunden), ist auch eine wichtige Voraussetzung für die Langzeitstabilität bei der Spektrenakkumulation erfüllt.

Es wurde zur Steuerung des Meßdurchlaufes sowie der Übernahme und Vorverarbeitung der Meßwerte das NMR-Spektrometer mit einem CAMAC-Mikrocomputer AMCA-80 on-line gekoppelt, der wiederum in das bereits vorhandene Lokale Computernetz ATLAS des ZfI integriert wurde, wodurch alle in dem lokalen Computernetz vorhandenen Ressourcen zur Spektrenauswertung im Realzeitbetrieb genutzt werden können.

2. Taxonomie

Die Kopplung des NMR-Spektrometer BS 487 C/TESLA mit einem CAMAC-Mikrocomputer AMCA-80 realisiert:

a) die Steuerung des Spektrometers zur Meßdatenerfassung,
b) die Übernahme der Meßdaten vom Spektrometer sowie eine Zwischenspeicherung im Mikrocomputer,
c) eine Meßdatenaufbereitung einschließlich Spektrenakkumulation,
d) die Kommunikation einschließlich Datenübertragung mit einer EDVA EC 1040,
e) die Ausgabe der im EC 1040 ermittelten Analysenergebnisse am Meßort.

Weitere Möglichkeiten ergeben sich durch die Integration in das Lokale Computernetz ATLAS des ZfI, in dem eine EDVA EC 1040 sowie unterschiedliche Prozeßrechnersysteme auf Mini- und Mikrorechnerbasis zusammengeschlossen sind. ATLAS ist ein Realzeitsystem für mehrere Forschungskollektive mit voneinander abweichenden Aufgabenprofilen sowie unterschiedlichen Meß- bzw. Experimentapparaturen [1, 3]. Die Prozeßrechner dienen zur Experimentsteuerung und Meßwertaufbereitung. Auswerteprogramme mit hohem Rechen- und großem residenten Speicherbedarf werden am EC 1040 abgearbeitet. Das System ermöglicht durch eine vom Nutzer gesteuerte interaktive Betriebsweise zwischen der Meßapparatur, dem Prozeßrechner und der ESER-EDVA, die Meßdatenerfassung, die Meßdatenaufbereitung und -verarbeitung einschließlich der Ergebnisausgabe am Ort der Meßapparatur während der Durchführung umfangreicher kernphysikalischer Messungen.

[1]) Anschrift: DDR-7050 Leipzig, Permoserstr. 15

In der EDVA EC 1040 erfolgt für die mit dem NMR-Spektrometer aufgenommenen Spektren:

a) die Isotopenanalyse,
b) die Linienformanalyse und
c) eine spezielle Spektrenakkumulation mit Synchronisation.

Diese Auswerteprogramme werden im Realtime-Regime abgearbeitet. Die gesamte Dialogführung, wie

- Auswahl der Programmroutinen im CAMAC-Mikrorechner,
- Eingabe von Parametern und Führungsgrößen,
- Auswahl der Auswerteverfahren und -routinen im EC 1040,
- Auslösen des Datentransfer zwischen Experimentcomputer und EC 1040,
- Reagieren im Fehlerfall,

erfolgt ausschließlich über die Bedieneinheit des CAMAC-Mikrocomputers.

3. Hardware

Die on-line-Koppling zwischen dem NMR-Spektrometer BS 487 C/TESLA und dem Mikrocomputer AMCA erfolgte auf der Basis des international standardisierten Instrumentierungssystems CAMAC unter Verwendung von Modulen, die — bis auf eine Ausnahme — im Zentralinstitut für Kernforschung Rossendorf der AdW der DDR hergestellt wurden (siehe Abb. 1).

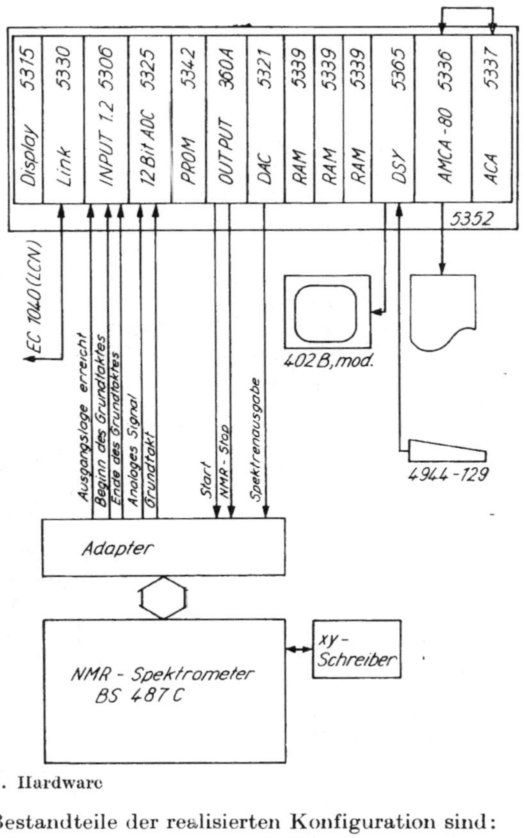

Abb. 1. Hardware

Die Bestandteile der realisierten Konfiguration sind:

5336	Intelligent Crate Controller AMCA-80
5337	Auxiliary Controller Adapter
5339	4 K RAM-Speichermodule
5342	16 K EPROM-Speichermodule
5306	Paralleles Daten-Eingabe-Register SI 1.2

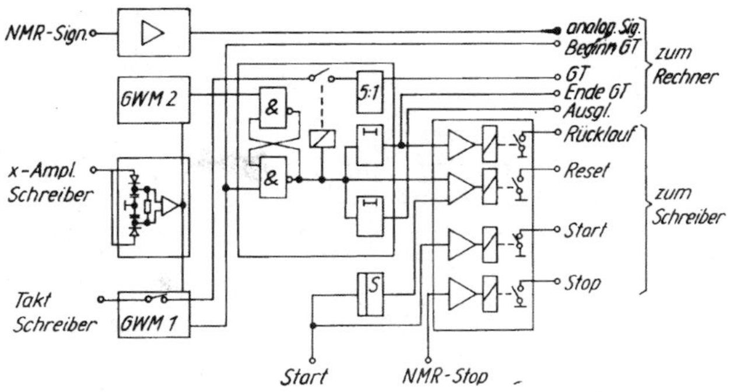

Abb. 2. Adapter

5325	Analog-Digital-Konverter 12 Bit
360 A	Paralleles Daten-Ausgabe-Register (POLON)
5321-10	Digital-Analog-Konverter 2 × 10 Bit
5330	Link
5365	Alphanumerischer Display Driver

Für den Anschluß des NMR-Spektrometers BS 487 C an CAMAC wurde ein spezieller Adapter (siehe Abb. 2) entwickelt, der folgende Forderungen erfüllt:

1) Lineare Verstärkung des NMR-Signalpegels um den Faktor 100 mittels des Verstärkerbausteines AdW-S 18. 4. 1 — 2.

2) Auswahl des Spektrenbereiches mittels zweier Grenzwertmelder (GWM) des Typs AdW-S 18. 4. 2 — 7, deren Eingangssignal durch Gleichrichtung der Impulsfolge am Linearpotentiometer des Schreibers und durch anschließende zehnfache Verstärkung gewonnen wird. Eine Pegelunterschreitung am GWM 1 liefert das Signal „Beginn des Grundtaktes", eine Pegelunterschreitung am GWM 2 bestimmt das Ende der Digitalisierung des Spektrums bei hohem Magnetfeld.

3) Bereitstellung des Grundtaktsignals aus dem Steuersignal für den Schrittmotor des Schreibers, wobei dem Spektrenbereich von 1000 Hz 2000 Grundtaktimpulse entsprechen sollen,

4) Interfacefunktionen für die Mikrocomputer-Steuerung des Spektrometers. Nach dem Startsignal vom Rechner liegt das analoge NMR-Signal am Ausgang des Adapters so lange an, bis GWM 2 umschaltet. Danach wird — ausgelöst von einer Speicherschaltung — über die Resetfunktion der NMR-Signaleingang an Masse gelegt und erst nach einer Verzögerung von 0,5 s mit dem Signal „Ende Grundtakt" die Digitalisierung beendet. Diese Maßnahme soll die Erkennung des Spektrenendes durch die Auswerteprogramme im Großrechner erleichtern. Mit dem Signal „Ende Grundtakt" beginnt automatisch der Rücklauf des Schreibers. Das Signal nach Pegelüberschreitung von GWM 1 löst nach einer Zeitvergrößerung von 40 ms das Signal „Ausgangslage erreicht" aus. Dieses Überfahren des GWM 1 soll sicherstellen, daß bei der nächsten Spektrenaufnahme zu Beginn der Digitalisierung mechanische Einschwingvorgänge nach dem Schreiberstart bereits abgeklungen sind.

4. Software

Für den realisierten Einsatzfall wurden folgende Softwarekomponenten entwickelt;

a) Software für Mikrocomputer AMCA-80,
b) Software des Lokalen Rechnernetzes ATLAS (siehe [1, 3].)
c) Auswerteprogramme für die EDVA EC 1040 [5].

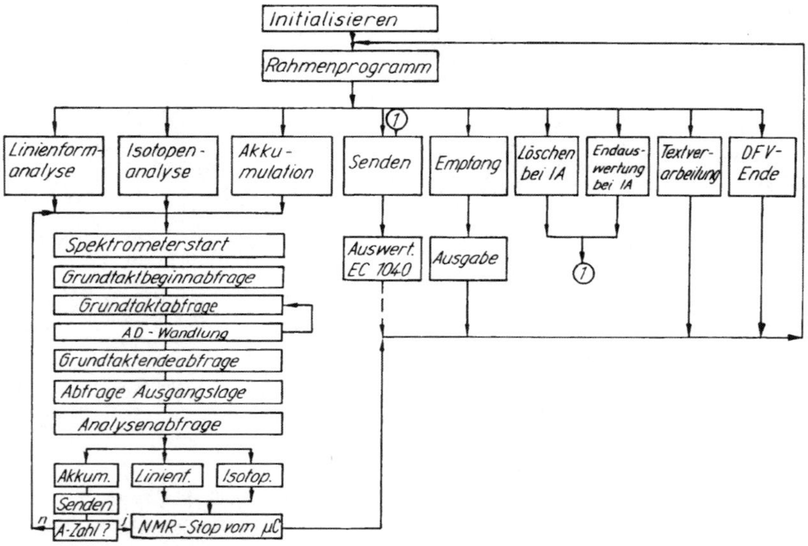

Abb. 3. Software für MC AMCA-80 (vereinfachte Darstellung)

4.1. *AMCA-80-Software*

Eine vereinfachte Darstellung der für den AMCA-80 entwickelten Programmroutinen zur Automatisierung der Meßwerterfassung und -verarbeitung des NMR-Spektrometers BS 487 C und dessen Integration in das Lokale Rechnernetz ATLAS zeigt Abb. 3.

Die gesamte Arbeit des AMCA-80 wird durch das Rahmenprogramm RAPRO koordiniert. Es aktiviert die Routinen zur automatisierten Spektrenaufnahme und dient zum Dialog des Nutzers mit dem AMCA-80 sowie dem EC 1040 durch Auswahl der Aktionen und der für die jeweiligen Aktionen benötigten Parameter. Einen Überblick über die programmierten Aktionen zeigt Tab. 1.

Tab. 1. Programmierte Aktionen am AMCA-80

A 1	Aufnahme der Spektrenwerte und senden zum EC 1040 bis Akkumulationszahl erreicht
A 2	Bereitstellung des auf AMCA-80 akkumulierten Spektrums, Senden zum EC 1040, Empfang der Ergebnisse
A 3	Spektrenakkumulation mit AMCA-80, Auswertung erfolgt über A 2
D	Beendigung der DFV-Arbeit
E	Empfang der Ergebnisse vom EC 1040
F	Fortsetzung der Akkumulation
H	Handdaten eingeben
I	Isotopenanalyse
L	Linienformanalyse
M	Rücksprung in Monitor
N	Neues Spektrum aufnehmen und abspeichern
O	Ein/Ausgabe der Spektrenwerte über logische Geräte
S	Senden zum EC 1040
T	Textausgabe auf dem Hardcopy-Gerät
X	Löschen des letzten Ergebnisses bei Isotopenanalyse
Y	Endauswertung der Isotopenanalyse
*	Unterdrücken des automatischen Übergangs vom Sende- zum Empfangsmodus bei A, S, X und Y

Das Rahmenprogramm fordert nach Ausführung einer Aktion jeweils den Nutzer auf, eine neue Aktion auszulösen. Dieser Aufruf kann vom Nutzer in beliebiger, jedoch vom jeweiligen Problem abhängigen, Reihenfolge zu verschiedenen Zeiten zur Steuerung der Arbeit des Systems aktiviert werden. Fehlbedienungen werden auf dem Bildschirm kommentiert, dann erfolgt eine neue Aufforderung. Zur Dialogführung wurden sowohl der freie als auch der geführte Dialog realisiert.

Beim freien Dialog gibt der Nutzer nach Anforderungen durch das System Anweisungen in einer speziellen Sprache ein, während beim geführten Dialog der Nutzer vom System aufgefordert wird, passende Eingaben auf bestimmte Fragen vorzunehmen.

Die für die Auswerteprogramme notwendigen Parameter sowie einige organisatorische Angaben, wie z. B. Datum und Spektrennummer, werden als Handdaten bezeichnet, deren Eingabe positionsorientiert erfolgt. Sie werden durch Leerzeichen getrennt, ihre Übernahme durch den Rechner erfolgt zeilenweise.

4.2. *Spektrenaufnahme*

Nach Abarbeiten der Aktionen A 1, A 3, I, L oder N (siehe Tab. 1) wird die Spektrenaufnahme mit einem speziellen Dialog eingeleitet. Dabei wird u. a. abgefragt, ob sich das Spektrometer in Ausgangslage befindet. Erst wenn dies der Fall ist, erfolgt der automatisierte Durchlauf des Spektrometers und die Übernahme der Meßwerte. Während der Spektrenaufnahme wird die aktuelle Anzahl der noch erforderlichen Akkumulationen ständig angezeigt.

Nach dem Beenden eines Meßdurchlaufs wird das Spektrometer in seine Ausgangslage zurückgebracht. Bei der Linienformanalyse und der Isotopenanalyse erfolgt ein Rücksprung in das Rahmenprogramm, und das Senden zum EC 1040 wird durch den Namen S gesondert ausgelöst. Bei der Akkumulation wird während der Rückwärtsbewegung das Spektrum zum EC 1040 gesendet und gleichzeitig auf dem AMCA akkumuliert. Mit Abschluß der Akkumulation werden entweder die Ergebnisse vom EC 1040 empfangen (A 1) oder es erfolgt ein Rücksprung in das Rahmenprogramm (+A 1). Mit Aufruf von A 2 wird das im AMCA akkumulierte Spektrum aufbereitet, zum EC 1040 gesendet und anschließend werden die Ergebnisse erwartet bzw. bei +A 2 erfolgt ein Rücksprung in das Rahmenprogramm. Nach Ausführen der Aktion A 2 steht im Handdatenpuffer eine „1" als Akkumulationszahl. Bei der Aktion A 3 wird die Spektrenakkumulation nur im AMCA-80 durchgeführt und zur Weiterbearbeitung ist A 2 zu aktivieren.

5. Ausgaben

Für die Darstellung der Ausgabeinformationen sind ein Display, ein Hardcopy-Gerät und ein XY-Schreiber vor-

Tab. 2. Allgemeine Fehlermitteilungen

*HANDDATEN FEHLERHAFT	
*ZEI – UEB	– Bei Handdateneingabe wurden mehr als 58 oder keine Zeichen pro Zeile eingegeben
*PUFFER LEER	– ein leerer Spektrenpuffer soll ausgegeben werden
*SPEKTROMETER IN AUSGANGSLAGE BRINGEN –	– beim Start der Spektrenaufnahme war Spektrometer nicht in Ausgangslage
*KEINE SIGNALE VOM SPEKTROMETER –	– keine Q-Antwort vom SI 1·2 Inputregister
*SPEKTRENAUFNAHME WIRD WIEDERHOLT –	– keine Q-Antwort vom ADC (falsche Spektrenwerte)
*X-ERROR	– keine X-Antwort bei CAMAC-Beefhlen
*Q-ERROR	– keine Q-Antwort bei CAMAC-Befehlen
*FALSCHE MARKE	– nicht def. Namen (Marke) eingegeben oder fehlerhafte Verwendung

Tab. 3. Nachrichten der ETOS-Steuer- und Verarbeitungsprogramme

*FEHLERHAFTE AUSGABEANFORDERUNG
*LAENGE DER HANDDATEN UNBEKANNT – CR NICHT GEFUNDEN
*PUFFER FUER HANDDATEN LEER
*UNGUELTIGE PARAMETERLAENGE FUER parameter
Folgende Parameter können auftreten:
VERARBEITUNGSART, AKKUMULATIONSZAHL, DATUM, SPEKTRENNUMMER, FREQUENZANFANG, PARAMETER, LOGISCHE VARIABLE, ABTASTZEIT, SUBSTANZNUMMER
*HANDDATEN UNVOLLSTAENDIG
*HANDDATEN UNVOLLSTAENDIG ODER CR NICHT GEFUNDEN
*UNGUELTIGE VERARBEITUNGSART
*UNGUELTIGE PARAMETERANGABE FUER parameter

Tab. 4. Sende- und Empfangsfehler

Empfangsfehler	
E1	kein ENQ bei Verbindungsaufnahme
E2	Erstes Zeichen ist kein STX oder EOT
E3	Kein ETX nach Text oder falsches LRC-Zeichen nach 2 Wiederholungen
E4	Kein STX bei Wiederholung
E5	Steuerzeichen im empfangenen Text
EØ	ENQ im empfangenen Text, Empfang wird wiederholt
Sendefehler	
S1	kein ENQ bei Verbindungsaufnahme
S2	kein EOT vor Senden Handdaten
S3	NAK nach Senden Handdaten (2 Wiederholungen)
S4	NAK nach Senden Spektrenwerte (2 Wiederholungen)
S5	kein EOT oder NAK nach Senden Spektrenwerte
S6	kein EOT oder NAK nach Senden Handdaten
SØ TO H	Zeitfehler beim Senden der Handdaten
SØ TO S	Zeitfehler beim Senden der Spektrenwerte

handen. Auf dem Display erscheinen grundsätzlich alle Ausgabeinformationen, während die durch die Auswerteprogramme im EC 1040 berechneten Ergebnisse (siehe [5]) zusätzlich über das Hardcopy-Gerät ausgegeben werden.

Die bei einem verteilten Informationsverarbeitungssystem mit interaktiver Betriebsweise gebotene Unmittelbarkeit, die gegenüber einem batchorientierten System die viel stärkere Integration eines solchen Systems in den Arbeitsablauf des Nutzers erlaubt, erforderte besonders den absoluten Ausschluß der Gefahr des Eintretens irgendwelcher Ausnahmesituationen im Systemverhalten. Deshalb war das Erreichen andauernder Kommunizierfähigkeit oberstes Ziel, um den für das Arbeiten des Systems unumgänglich erforderlichen Nachrichtenverkehr für die Dauer einer Meßserie unter allen Umständen aufrecht erhalten zu können. Bei erkannten Fehlabläufen oder fehlerhaften Eingaben werden differenzierte Fehlermitteilungen ausgegeben. Dabei wird unterschieden zwischen allgemeinen Fehlermitteilungen, die vom AMCA-80 erkannt werden (siehe Tab. 2) und Nachrichten der Steuer- und Verarbeitungsprogramme des Teleprocessing-Monitors ETOS in der EDVA EC 1040 (siehe Tab. 3). ETOS wurde für seine Funktion im Lokalen Rechnernetz ATLAS um wesentliche Funktionen erweitert. Wird bei der Übertragung ein nicht der Datenübertragungsprozedur [1] entsprechendes Zeichen empfangen, dann werden eine Fehlernummer und das empfangene Zeichen (Hexadezimal) ausgegeben (siehe Tab. 4). Danach erfolgt der Rücksprung in das Rahmenprogramm und mit der erneuten Eingabe des Namens (Marke) kann die Arbeit fortgesetzt werden.

Bei der Abarbeitung der Programme Spektrenaufnahme, Senden, Empfangen oder Beendigung der DFV-Arbeit wird stets der aktuelle Bearbeitungsstand auf dem Bildschirm in Form eines „Blinkenden Zeichens" angezeigt (Statusanzeige, siehe Tab. 5).

Tab. 5. Statusanzeigen

Spektrenaufnahme:	
'A'	Warten auf Grundtakt-Anfang
'W'	Warten auf Grundtakt-Ende
'G'	Meßwerte werden übernommen
'R'	Warten auf Ausgangslage
Senden:	
'V'	Verbindungsaufnahme, Warten auf ENQ
'O'	Warten auf EOT
'H'	Senden Handdaten
'S'	Senden Spektrenwerte
Empfang:	
'V'	Verbindungsaufnahme, Warten auf ENQ
'T'	Warten auf STX
'E'	Empfang
Beendigung DFV:	
'V'	Verbindungsaufnahme, Warten auf ENQ
'O'	Warten auf EOT

6. Zusammenfassung

Es wird die on-line-Kopplung für ein cw-NMR-Spektrometer (BS 487 C/TESLA) beschrieben, die mittels der CAMAC-Mikrorechnertechnik die Meßdatenauswertung an einer EDVA ES 1040 in Dialogbetrieb unter Realzeitbedingungen realisiert. Die Kommunikation erfolgt über ein Lokales Computernetz.

Mit diesem Rationalisierungsvorhaben wurde das Ziel verfolgt, die Empfindlichkeit des Gerätes durch Nutzung der Spektrenakkumulation zu erhöhen, den Zeitaufwand für die Isotopenanalytik zu senken und durch Linienformanalysen exakte NMR-Parameter bestrahlter Modellmembranen zu erhalten. Das System wird seit Anfang 1983 routinemäßig genutzt.

Der vorliegende Beitrag gibt einen Überblick über die Hardware des Systems, den Datentransfer, die Mikrorechnersteuerung und die Spektrometersteuerung. Die Vorstellung der Auswerteprogramme am ES 1040 mit Darstellung der erzielten Ergebnisse wird in einem gesonderten Beitrag behandelt [5].

Eingegangen am 25. 2. 1985

Literatur

[1] *B. Lindemann, R. Radmanesh,* Isotopenpraxis *18* (1982) 96
[2] *H. Sprinz, E. Winkler, H. Schäfer,* Isotopenpraxis *17* (1981) 171
[3] *B. Lindemann,* Verteilte Datenverarbeitungssysteme – Verarbeitungsprinzipien und Kommunikationssysteme. INFO 84 Dresden 6.–10. Februar 1984, S. 88
[4] *B. Lindemann,* Nachrichtentechnik/Elektronik *34* (1984) 326
[5] *E. Winkler* u. a., Isotopenpraxis (in Vorbereitung)

Kurze Originalmitteilungen

Isotopenpraxis 22 (1986) 1, S. 36 bis 37

Technetiumverbindungen als Katalysatoren für die Reaktion von Hydroperoxiden mit Olefinen

S. Rummel, B. Lorenz

Zentralinstitut für Isotopen- und Strahlenforschung der AdW der DDR)[1]

Keywords

alkenes; catalysis; epoxides; technetium complexes; technetium compounds

In den letzten Jahren sind eine Reihe katalytischer Eigenschaften von Technetium und seinen Verbindungen aufgefunden worden [1 – 3]. Wir untersuchten mehrere Komplexverbindungen des Technetiums auf ihre katalytische Wirksamkeit bei der Reaktion von Olefinen mit Hydroperoxiden (*Halcon*-Reaktion), die ohne Katalysator nicht abläuft (Gl. 1) [4]:

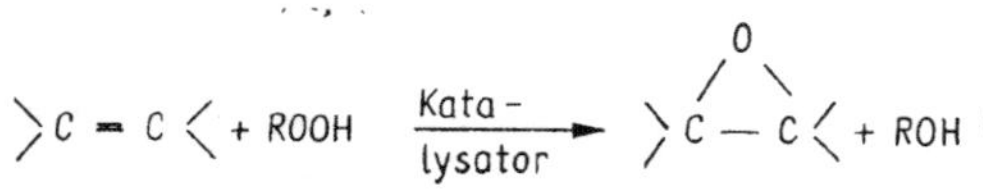

(1)

Bei der Reaktion von Cyclohexen mit tert.-Butylhydroperoxid (t-BuOOH) wurden in Anwesenheit von Technetiumkomplexen Ausbeuten an Epoxycyclohexan bis zu 32%, bezogen auf das umgesetzte Hydroperoxid, erhalten. Konkurrierend zu der polaren Umsetzung des Hydroperoxids in Epoxide katalysieren Technetiumverbindungen auch die homolytische Zersetzung des t-BuOOH unter Bildung freier Radikale [4] (Gl. 2 + 3). Ferner fanden wir [5, 6], daß Technetiumverbindungen auch eine Weiterreaktion des gebildeten Epoxids in radikalinduzierten Reaktionen katalysieren, so daß aus den Epoxidausbeuten nicht direkt auf die katalytische Aktivität der Technetiumkomplexe in der Reaktion (Gl. 1) geschlossen werden kann.

$$\text{t-BuOOH} + M^{n+} \rightarrow \text{t-BuOO}\cdot + H^+ + M^{(n-1)+} \quad (2)$$

$$\text{t-BuOOH} + M^{(n-1)+} \rightarrow \text{t-BuO}\cdot + OH^- + M^{n+} \quad (3)$$

Wir untersuchten deshalb in der Reaktion von t-BuOOH mit Cyclohexen den Einfluß des Radikalfängers 4-Methyl-2,6-di-tert.butylphenol (Jonol) auf die Ausbeuten an Epoxycyclohexan (Tab. 1). Als Katalysatoren wurden eine Palette von Komplexverbindungen des Technetiums eingesetzt, die von uns nach Literaturvorschriften erhalten bzw. eigens dargestellt wurden (VII, VIII, X, XII). Zum Vergleich wurden Komplexverbindungen von Mo und Mn verwendet, die nach Literaturvorschriften hergestellt wurden. Die Umsetzung erfolgte in geschlossenen Ampullen über einen Zeitraum von 10 h bei 80 °C unter Stickstoff.

Tab. 1. Einfluß eines Inhibitors auf die Ausbeute an Epoxycylohexan[a])

Katalysator[b])		ohne Jonol		mit Jonol	
		t-BuOOH[c])	Epoxid[d])	t-BuOOH[c])	Epoxid[d])
$TcCl_4(PPh_3)_2$	I	100	14·9	100	31·3
$TcCl(acac)(PPh_3)_2$	II	100	11·9	100	24·2
$Tc_2(CO)_{10}$	III	100	31·4	100	48·4
$TcCl_4(PhEt_2P)_2$	IV	99	9·5	100	22·7
$TcCl_3(Me_2PhP)_3$	V	100	5·5	100	17·7
$[TcCl_6][PPh_3H]_2$	VI	100	7·8	100	25·8
$KTc_2(ac)_4Cl_2$	VII	96	6·6	25	13·3
$[TBA][TcBr_4NO]$	VIII	60	5·9	100	9·9
$TcCl_4(Ph_2EtP)_2$	IX	100	14·3	100	26·5
$TcN(dtc)_2$	X	100	9·4	100	–
$TcBr_4(PPh_3)_2$	XI	100	9·6	100	2·4
$[TcBr_6][PPh_3H]_2$	XII	100	11·1	100	–
$TcCl(CO)_3(PPh_3)_2$	XIII	100	12·8	98	–
$Mo(CO)_6$		96	99·0	100	85·0
$MoO_2(acac)_2$		94	94·2	95	79·7
$Mn(acac)_3$		78	5·7	80·2	1·5

[a]) Reaktionsbedingungen: 120 ml Cyclohexen, 60 ml Benzen und 4,2 ml Dichlorbenzen ergeben eine Stammlösung; 1,5 ml Stammlösung, 5 × 10^{-6} mol Katalysator, 5 × 10^{-4} mol t-BuOOH, 5 × 10^{-4} mol Jonol, 80 °C, 10 h Reaktionszeit
[b]) acac = Acetylacetonat; Ph = Phenyl; dtc = Diethyldithiocarbamat; TBA = Tetrabutylammonium; Et = Ethyl; ac = acetat; Me = Methyl
[c]) Umsatz von t-BuOOH
[d]) Ausbeute an Epoxycyclohexan in % bezogen auf 100 % des eingesetzten HPO

Die Ergebnisse in Tab. 1 zeigen, daß der Zusatz von Jonol in Abhängigkeit vom eingesetzten Katalysator sowohl eine Erhöhung wie auch eine Erniedrigung der Ausbeute an Epoxycyclohexan bewirken kann.

[1]) Anschrift: DDR-7050 Leipzig, Permoserstraße 15

Bei der Reaktion mit Molybdän-Verbindungen als Katalysator wird entsprechend den Vorstellungen über den Katalysemechanismus [7, 8] unter den gewählten Bedingungen ein fast 100%iger Umsatz des eingesetzten Hydroperoxids gemäß Gl. (1) beobachtet. Durch den Zusatz von Jonol wird die Epoxidausbeute verringert, wahrscheinlich durch die Weiterreaktion des gebildeten Epoxids in einer nucleophilen Ringöffnungsreaktion mit dem Jonol, die durch Mo-Verbindungen katalysiert wird [9].

Umgekehrt steigen die Epoxidausbeuten in Anwesenheit von Jonol bei der Katalyse durch die Mehrzahl der eingesetzten Tc-Verbindungen (I – IX). Auch hier können nucleophile Epoxidringöffnungsreaktionen mit Jonol die Epoxidausbeuten negativ beeinflussen, jedoch erwiesen sich Tc-Verbindungen in dieser Reaktion als wenig aktiv [6]. Offensichtlich überwiegt hier der Einfluß von Jonol als Radikalfänger. Durch das Abfangen freier Radikale aus der katalytischen Zersetzung des t-BuOOH (Gl. 2 + 3) werden die radikalinduzierte Zersetzung des Hydroperoxids [1] wie auch Tc-katalysierte radikalische Epoxidringöffnungsreaktionen unterdrückt, die zur Bildung von Peroxyketonen und polymeren Epoxiden führt [6]. Folgerichtig kommt es durch die Hemmung der Radikalreaktionen zu einer Erhöhung der Epoxidausbeute.

Unter den von Jonol unterdrückten Nebenreaktionen des tert.-Butylperoxidradikals (Gl. 2) befindet sich auch die Reaktion (Gl. 4), die zur Bildung von Epoxycyclohexan auf radikalischem Wege führt [11]:

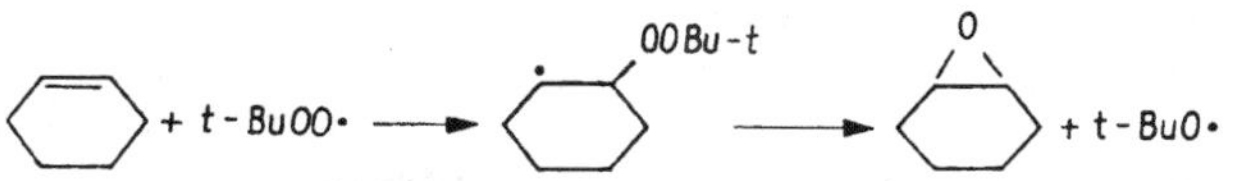

(4)

Die Verringerung bzw. das völlige Verschwinden von Epoxycyclohexan unter den Reaktionsprodukten in Anwesenheit von Jonol bei der Katalyse durch die Tc-Komplexe X bis XIII bzw. durch $Mn(acac)_3$ (Tab. 1) deuten auf einen beträchtlichen Anteil der radikalischen Epoxidbildungsreaktion (Gl. 4) an der Epoxidausbeute in diesem Reaktionssystemen hin.

Eingegangen am 15. 5. 1985

Literatur

[1] *K. Schwochau*, Radiochim. Acta *32* (1983) 139
[2] *B. Bayerl, M. Wahren*, Z. Chem. *21* (1981) 149
[3] *B. Lorenz, H. Rommel, M. Wahren*, Z. Chem. *22* (1982) 224
[4] *S. Rummel, D. Schnurpfeil*, Oxid. Comm. *6* (1984) 319
[5] *S. Rummel, D. Schnurpfeil, L. Willecke*, Z. Chem., *25* (1985) 150
[6] *S. Rummel, D. Schnurpfeil*, J. prakt. Chem., im Druck
[7] *R. A. Sheldon*, J. Mol. Catal. *7*, (1980) 149
[8] *J. Sobcak, J. J. Ziolkowski*, J. Mol. Catal. *13* (1981) 11
[9] *G. Lauterbach, G. Posselt, R. Schäfer, D. Schnurpfeil*, J. prakt. Chem. *323* (1981) 101
[10] *E. S. Gould, R.R. Hiatt, K. C. Irwin*, J. Amer. Chem. Soc. *90* (1968) 4573
[11] *R. A. Sheldon, J. A. van Doorn*, J. Catal. *31* (1973) 427

Isotopenpraxis 22 (1986) 1, S. 37 bis 40

Buchbesprechungen

Nuclear Techniques in Tropical Animal Diseases and Nutritional Disorders
Proc. of a Consultants Meeting on the Application of Nuclear Techniques in the Study of Tropical Animal Diseases Organized by the Joint FAO/IAEA Division of Isotope and Radiation Applications of Atomic Energy for Food and Agricultural Development and held in Vienna from 13 to 16 June 1983

Vienna: IAEA 1984. 212 S., 21 Abb., 29 Tab., 724 Lit. (Proc. Series STI/PUB/675); kart. ÖS 420.–

Im Weltmaßstab verenden jährlich mehr als 50 Mio Rinder sowie 100 Mio Schafe und Ziegen an krankheitsbedingten Störfaktoren. Diese sind wesentliche, die Landwirtschaftsentwicklung bedrohende Faktoren in vielen Teilen der Welt, insbesondere in jungen Nationalstaaten, und führen damit zu einer bedrohenden Einschränkung der menschlichen Ernährungsgrundlage. Überwiegend sind hiervon die tropischen und subtropischen Regionen betroffen. Zahlreiche Tiererkrankungen werden auf polyfaktorielle Störungen im Reproduktionsmangagement zurückgeführt. Hierbei spielen als komplikative Faktoren Infektionskrankheiten (Parasiten, Bakterien, Viren) und Mangelernährung eine besondere Rolle.

Der Einsatz der Nukleartechnik in der Ursachenermittlung, Diagnostik, teilweise auch in der Bekämpfung dieser Störfaktoren war Gegenstand des Treffens einer Konsultantengruppe der FAO/IAEA, deren 44 Beiträge von namhaften Autoren im vorliegenden Band jeweils mit Abstracts, Schlußfolgerungen und Literatur enthalten sind. Sie lassen sich folgenden Problemkreisen zuordnen: Immundiagnostik bei Virusinfektionen und Parasitosen – Störungen im Reproduktionsmanagement – Mengenelementmetabolismus, Dysfunktion, Toxizität – Immunkompetenz und Immunpathogenese – Parasitäre Infektionen (Nematoden des Gastrointestinaltraktes, Babesiose, Theileriose). Aus parasitologischer Sicht wird auf den Einsatz von Radiovakzinen zur Immunprophylaxe besonders bei protozoären Infektionen, wie Babesiose (*Wright*, Queensland/Australien) und Theileriose (*Irvin*, Nairobi/Kenya) orientiert und damit der internationale Trend herausgestellt. Die Beiträge enthalten zahlreiche Detailangaben zu methodischen Fragen, wie z. B. Energiedosen bei Radiovakzinen (Babesiose, Theileriose, Anaplasmose), Antigenpräparationen, Progesteron-Radioimmunoassay (P_4–RIA), Immunologische Technik mit Isotopenmarkern.

Aus der Sicht der 5 Sektionen (Immunologie, Infektionskrankheiten, Parasitäre Erkrankungen, Reproduktionsstörungen, Ernährungsstörungen) werden abschließend eine Einschätzung der gegenwärtigen Situation und Empfehlungen zum Einsatz der Nukleartechnik auf den jeweiligen Gebieten gegeben. Die Nukleartechnik gestattet in Verbindung mit anderen Standardmethoden eine Erweiterung der Untersuchungsmöglichkeiten und damit einen tieferen Einblick in die Pathogenese krankheitsbedingter Störfaktoren. Der Band ist für Virologen, Parasitologen, Immunologen ebenso aktuell wie für Zuchthygieniker und Ernährungswissenschaftler.

R. Jungmann

Handbuch der medizinischen Radiologie
Band XV, Nuklearmedizin
Teil 3, Diagnostik II, Pädiatrische Nuklearmedizin

Von 21 Fachwissenschaftlern; redigiert von H. Hundeshagen. Berlin, Heidelberg, New York, Tokyo: Springer 1985. 604 S. (157 in Engl.), 182 Abb. mit 362 Einzeldarstellungen (einige in Farbe), Namenverzeichnis, Sachverzeichnis (deutsch/engl.); geb. DM 790,–

Vom Handbuch der medizinischen Radiologie (insgesamt 20 Bände umfassend) liegt nunmehr der Teil 3 des Bandes XV, Nuklearmedizin, vor. Er enthält in mehreren Kapiteln zusammenfassende Darstellungen über solche Gebiete der klinischen Nuklearmedizin, die sich in den letzten Jahren besonders schnell weiterentwickelt haben. Dies trifft in ausgeprägtem Maße für die Herz- und Kreislaufdiagnostik zu, gilt aber auch für die *in-vitro*-Diagnostik, die nuklearmedizinische Tumordiagnostik und die pädiatrische Nuklearmedizin.

Den meisten Raum (362 S.) nimmt die Herz- und Kreislaufdiagnostik ein, die sich zu einem diagnostisch besonders wichtigen Gebiet der Nuklearmedizin entwickelt hat. Da die relevanten Untersuchungsverfahren zu den schwierigsten und aufwendigsten innerhalb der Nuklearmedizin gehören, ist dies voll und ganz gerechtfertigt und ermöglicht es dem Leser, an dem breiten Erfahrungsschatz der Bearbeiter zu partizipieren. Nach einer kurzen Einführung (*Hundeshagen*) werden die Verfahren im einzelnen dargestellt: Radiokardiographie mit der Multikristall-γ-Kamera (*Schad*), Herzbinnenraum-Szintigraphie (*Adam* und *Bitter*), Untersuchung des Myokard-Blutflusses (*Lichtlen, Engel, Hundeshagen*), Myokard-Szintigraphie mit ^{201}Tl (*Pretschner*), Darstellung von akuten Myokard-Infarkten (*Parkey, Lewis, Bonte, Buja, Willerson*), Shunt-Diagnostik (*Markisz, Kuruc, Treves*), Untersuchungsverfahren bei peripheren Durchblutungsstörungen und der arteriellen Verschlußkrankheit (*Alexander*) sowie Diagnostik von Thromben und Emboli (*Buttermann, Pabst*). Der Informationswert der Darstellungen ist außerordentlich hoch. Jedes Kapitel enthält ausführliche Angaben über Radiopharmaka, Applikation, Untersuchungs- und Auswertetechniken, Meß- bzw. Geräteaufwand, diagnostische Wertigkeit und Anwendungen. Zahlreiche Abbildungen, darunter außerordentlich gute Szintigramme bzw. Kameraaufnahmen, unterstützen das Anliegen der Verfasser wirkungsvoll. Der interessierte Leser verspürt noch die Dynamik der Entwicklung während des letzten Jahrzehnts auf diesem wichtigen Gebiet der nuklearmedizinischen Diagnostik, getragen von neuen Radiopharmazeutika und leistungsfähigen Kamera-Computer-Systemen.

Das Kapitel *in-vitro*-Diagnostik im nuklearmedizinischen Laboratorium (43 S.; *Wood*) bringt in gedrängter Form einen instruktiven Einblick in dieses wichtige Gebiet der Mikroanalytik. Im ersten Teil wird der Leser mit den theoretischen Grundlagen und der Arbeitsweise einschließlich Auswertung und Qualitätskontrolle der Radioligandassays, speziell des Radioimmunoassays, vertraut gemacht. Im zweiten Abschnitt werden die wichtigsten Anwendungsgebiete beleuchtet: Schilddrüse, Nebenniere, Reproduktion, Diabetes u. a. Trotz der gebotenen Kürze erhält der Leser einen guten Eindruck von der enormen Leistungsfähigkeit dieser Analysenverfahren, die aus der nuklearmedizinischen Diagnostik nicht mehr wegzudenken sind.

Ein weiteres Kapitel (58 S.) ist der Tumordiagnostik gewidmet (*Winkler*). Obgleich dieses Gebiet noch in vollem Fluß ist, sind doch in den letzten Jahren durch die Einführung nuklearmedizinischer Methoden in die klinische Onkologie die Möglichkeiten des Tumornachweises und der Verlaufskontrolle wesentlich erweitert worden. Auch hier werden nach einer Einführung in die allgemeinen Prinzipien des nuklearmedizinischen Tumornachweises zunächst die schon erprobten radioaktiven Testsubstanzen besprochen. Dann wird der Leser mit den Möglichkeiten ihres Einsatzes in der Lokalisationsdiagnostik bei zahlreichen Organen vertraut gemacht, wobei auch hier die Abbildungen wesentlich zur Information beitragen. Ein spezieller Abschnitt dieses Kapitels ist schließlich den Tumormarkern gewidmet.

Zum Schluß des Bandes kommt die pädiatrische Nuklearmedizin zu Wort (22 S.). Die Bearbeiter (*Ebel, Treves*) verweisen eingehend auf die Besonderheiten der Anwendung von Radiopharmaka beim kranken Kind. Das zweckmäßige diagnostische Vorgehen, z. B. zur Szintigraphie einzelner Organe, wird beschrieben, wobei auch auf die Anwendung alternativer bildgebender Verfahren hingewiesen wird.

Jedem Kapitel ist eine ausführliche, z. T. bis 1983 reichende Literatursammlung beigegeben. Das Werk bietet jedem, der sich in diese modernen diagnostischen Verfahren einarbeiten will, eine vorzügliche Möglichkeit. Druck und Ausstattung des Werkes sind hervorragend. *H. Herzmann*

Baumlos in die Zukunft?

D. Guratzsch (Hrsg.). München: Kindler Verlag 1984. 272 S., zahlr. Abb. u. Tab.; Ppb. DM 19,80

Man erwartet auf die provozierend gestellte Frage des Buchtitels eine Antwort, auf den nur schwer faßbaren Begriff des Waldsterbens, des rasch voranschreitenden und sich kontinental ausbreitenden Prozesses, der zur Zerstörung unserer Wälder führt. Der Herausgeber läßt 13 Wissenschaftler zu Wort kommen, die aus unterschiedlichen Blickwinkeln versuchen, sich dem komplexen und komplizierten Erscheinungsbild zu nähern und erklärende Antworten – oder Teilantworten – zu geben. Neben dem Erleben und den Beschreibungen des Tatbestandes versuchen zwei Autoren Erklärungen zu geben, die sich nicht ohne weiteres in Einklang bringen lassen. Der eine Autor – Forstoberrat Dr. *K. Thiele* – sieht die Hauptursache sozusagen in einer Reihe „oberirdischer" Stressoren (Hitze und Kälte, Schadinsekten, waldbauliche Fehler, Luftschadstoffe, um nur die wichtigsten zu nennen), während im Gegensatz dazu der Bodenkundler Professor Dr. *B. Ulrich* die Stressoren in die „unterirdische" Bodenzone verlagert, die sich als Folge der Luftschadstoffbelastungen irreversibel verändert hat und verändert (pH-Wert-Erniedrigung und damit Freisetzung von Mg, Al und Schwermetallen, biotische Schädigung der Wurzelsysteme, um auch hier einige zu nennen). Wenn auch manches heute noch nicht verständlich erscheint, so ist es auf Teilgebieten zumindest erklärbar, und ein gemeinsamer Nenner für alle Ursachen ist gegenwärtig noch einladend groß – ein ungelöster Problemberg liegt noch vor uns.

Es ist dem Herausgeber zu danken, daß eine Schrift vorgelegt werden konnte, die eine Zwischenbilanz darstellt und manche unkritische Diskussion über das Erscheinungsbild des Waldsterbens verhindern hilft. Herabsetzung des Schadstoffausstoßes und damit Herabsetzung der flächenspezifischen Schadstoffdeposition ist noch die wirksamste Maßnahme für den Wald und für die Menschen; denn: „Stirbt der Wald, so stirbt auch der Mensch" — wie es zum Schluß als Resümee in diesem Buch heißt. Nur weitere Forschung, waldbauliche Maßnahmen und energiepolitische Entscheidungen können dieses kontinentale Problem lösen und eine Zukunft für den Wald und die Menschen sichern.

H. Hübner

Mit dem Salz durch die Jahrtausende

H.-H. Emons, H.-H. Walter. Leipzig: VEB Deutscher Verlag für Grundstoffindustrie 1984. 227 S., 190 Abb. (23 farbig), 3 Tab., 142 Lit.; Leinen M 30,—

Das bei überwiegend vegetarischer Ernährung physiologisch bedingte unwiderstehliche Verlangen nach Kochsalz ist der tiefere Grund für die überragende Rolle, die das Natriumchlorid als Nahrungsmittelzusatz seit Jahrtausenden spielt, und ist zugleich die Ursache für die tiefe Verwurzelung des Salzes im Leben und Denken der Menschen insbesondere im Altertum und im Mittelalter. Heute zwar immer noch unentbehrlicher Nahrungsmittelzusatz, überwiegt — gemessen am Verbrauch — bei weitem die Rolle des Kochsalzes als Rohstoff für die chemische Industrie.

Die Autoren, zwei technisch orientierte Chemiker, die sich in ihrer Forschungsarbeit seit vielen Jahren mit der Stoffklasse der anorganischen Salze und auch mit dem Natriumchlorid im engeren Sinne beschäftigen, zeichnen in Verbindung mit der Entwicklung der menschlichen Gesellschaft ein plastisches Bild der Geschichte der Gewinnungsmethoden des Salzes aus Sole, Meerwasser und auf bergmännischem Wege sowie seiner vielschichtigen Nutzung. Durch eine Gliederung in sechs Kapitel, denen die Zeitepochen Urgeschichte, Antike, Mittelalter sowie die Zeitabschnitte 1500—1800, 1800—1900 und Gegenwart zugeordnet werden, schaffen sich die Verfasser den äußeren Rahmen für die sehr schwierige aber auch verlockende Aufgabe, die Historie einer der einfachsten chemischen Verbindungen über einen Zeitraum von mehr als 5000 Jahren zu verfolgen.

Im Abschnitt zur Urgeschichte werden die als Briquetage-Technik bezeichnete Salzgewinnung aus Solen mittels Gerätschaften aus Ton sowie der Salzbergbau in der Bronze- und Eisenzeit beschrieben. Weiterhin werden aus ethnografischen Berichten über Volksstämme, die noch auf der Stufe der Urgesellschaft leben, interessante Rückschlüsse auf Salzgewinnungsmethoden aus Ersatzstoffen gezogen. In den Kapiteln zur Salzgeschichte in der Antike und im Mittelalter wird neben einer Darstellung der Gewinnungsmethoden, der Organisation der Salzproduktion und der Entwicklung des Salzhandels vor allem gezeigt, daß das Salz die Gedanken- und Gefühlswelt in diesen geschichtlichen Epochen in erstaunlicher Weise beeinflußte und Eingang in Sitten und Gebräuche sowie in die Heilkunde aber auch in Religion und Aberglauben fand.

In den Zeitabschnitt 1500—1800 fällt die Errichtung von Gradieranlagen als wirksame technische Neuerung zur Anreicherung der Solen durch Verdunstung eines Teils des Wassers an der Luft, was im Prinzip allen Salinen die Versiedung konzentrierter Sole ermöglichte. 1791 wurde in Frankreich die erste Fabrik in Betrieb genommen, die Salz in Soda umwandelte und nach 1800 zur Errichtung zahlreicher Sodafabriken in Europa führte, wobei zunächst nach dem Verfahren von *Leblanc*, später nach *Solvay* produziert wurde.

Zu diesem Zeitpunkt setzt, wie die Autoren im fünften Kapitel zeigen, die Entwicklung des Natriumchlorids von einem Hilfsstoff für Gewerbezweige des Altertums und Mittelalters zu einem unentbehrlichen Rohstoff für die chemische Industrie ein, die heute in einer Schlüsselstellung bei der Herstellung von Chlor, Soda, Alkalilaugen und Natrium zum Ausdruck kommt. Die chemische Industrie verbraucht gegenwärtig etwa 120 Mio. t Salz im Jahr, von denen ca. $^2/_3$ in modernen Salzbergwerken und $^1/_3$ in Meeressalinen gewonnen werden.

Die Verfasser beenden ihren Streifzug durch die Geschichte des Salzes mit einer kurzen Abhandlung über das blaue Steinsalz. Auch wenn unter Physikern und Chemikern heute die Meinung vorherrsche, daß vorwiegend die Strahlung des Kalium-40 die auslösende Ursache für die Entstehung der Farbzentren (Gitterfehlstellen) im blauen Steinsalz sei, bleiben dabei, so die Autoren, viele Fragen offen, insbesondere zum räumlich eng benachbarten Auftreten von weißem und blauem Steinsalz sowie zum meist fehlenden Zusammenhang zwischen Kaliumgehalt und Blaufärbung.

H. D. Schmiedl

A Handbook of Radioactivity Measurements Procedures

Bethesda: National Council on Radiation Protection and Measurements 1985. 2. Aufl.; 592 S.; zahlr. Abb. u. Tab., ca. 760 Lit. (NCRP Report No. 58); geb. US $ 22,—

Das vom wissenschaftlichen Komitee für Radioaktivitätsstandards und -messungen des NCRP der USA nach der inzwischen vergriffenen 78er Auflage erneut und in erweiterter Form publizierte Buch ist in folgende Kapitel gegliedert: 1. Einführung, 2. Physik der Strahlungsdetektoren, 3. direkte (absolute) Aktivitätsmessung, 4. indirekte bzw. vergleichende Messung, 5. Präparationstechnik für Standardquellen, 6. Radioaktivitätsbestimmung und Identifizierung von Radionukliden in Labors für Umweltuntersuchungen, Medizin und Industrie, 7. Statistik und 8. Zuverlässigkeit, Standards, Meßvergleiche und Unsicherheitsangaben. Etwa das letzte Drittel des Buches umfaßt die Anhänge, deren wichtigster aus einer Tabelle von über 200 der für medizinisch-biologische und industrielle Anwendungen bedeutsamsten Radionuklide besteht. Sie ist die Wiedergabe von Rechnerausdrucken der Oak-Ridge Datenbanken mit Halbwertszeiten, Zerfallsarten, -häufigkeiten und -energien sowie Zeitintegralen über die mittleren Emissionsenergien. Fehlerangaben sind enthalten. Der letzte Teil der Anhänge betrifft die Zerfallsstatistik.

Seinem Charakter nach steht das Werk zwischen Lehrbuch und Monografie. Es ist sehr praxisfreundlich angelegt, bestens verständlich und dabei dennoch fast durchweg sehr exakt formuliert. Der Text wird durch nicht allzu viele, aber instruktive Abbildungen ergänzt. Nach Auffassung des Rezensenten hat das Buch seine größten Stärken bei der Darstellung der indirekten oder relativen Aktivitätsmessungen, die sowohl für die Praxis am wichtigsten sein dürften und bei denen sich — man denke insbesondere an die spektrometrischen Techniken — in den letzten Jahrzehnten auch die wesentlichsten Entwicklungen vollzogen haben.

Bestimmte Sachverhalte, z.B. Elementarprozesse der Wechselwirkung von Gammastrahlung mit Stoffen, auch die Grundlagen der Zerfalls- und Zählstatistik, werden in dem Band mehrfach und an verschiedenen Stellen angesprochen. Dadurch erhält es stellenweise überflüssige Redundanz (etwa Abschn. 4.2.2.). Dieses Manko an wechselseitiger Abstimmung ist ja häufig bei Werken mit einer ganzen Anzahl von Autoren und ließe sich hier am besten vermeiden, indem in Zukunft und in konzentrierter Form ein Kapitel „Strahlungswechselwirkung mit Stoffen" an den Anfang gestellt würde, das eben genau die Dinge enthalten sollte, auf die man bei der Strahlungsmeßtechnik zurückgreifen muß. Dann würden in den weiteren Kapiteln kurze Verweise ebenso genügen wie in Fällen der Statistik, die, wenn man ihr schon ein ganzes Kapitel einräumt, dann z.B. nicht noch einmal in einem Anhang zu erscheinen brauchte. Damit ließe sich Platz sparen, den man für Themen verwenden könnte, die man leider vergeblich sucht: So wird die Teilchenwechselwirkung kaum erwähnt, obwohl Reichweiten und Bremsvermögen für (Sekundär-) Elektronen im Zusammenhang z.B. mit Messungen der absorbierten Dosis über die Hohlraumionisation (das alte *Bragg-Gray*sche Prinzip!) elementare Begriffe sind. Die Strahlenschwächung in den Umhüllungsmaterialien der Detektionsmedien wird generell stiefmütterlich behandelt: Man vermißt sie z.B. bei der Besprechung der Szintillatoreffektivität (Abschn. 4.2.6.1.) völlig, während sie für Ionisationskammern in Tab. 8, S. 132, nur sehr pauschal und beziehungslos mitgeteilt wird.

Auf die Präparation von Standards entfallen nur 20 Seiten mit teils lockerer, kaum wertender Aufzählung; davon weniger als die Hälfte auf feste Präparate. Flächenpräparate sind überhaupt nicht berücksichtigt, obwohl sie als Standards für Abbildungsverfahren in der Nuklearmedizin bedeutsam sind. Autoradiografie und ihre ebenso notwendigen wie nützlichen Eichverfahren spielen in dem Band ebenfalls keine Rolle. Dem Rezensenten will scheinen, daß sich in Tab. 17, S. 300, ein grob verfälschender Druckfehler eingeschlichen hat: Ionisationskammern sind sicherlich nicht nur für den μCi-Bereich zu gebrauchen, wie dort angegeben, sondern ermöglichen leicht die Graduierung viel höherer Aktivitäten.

Trotz seiner Lücken und Disproportionen handelt es sich ohne jeden Zweifel um ein überaus nützliches Buch. Physiker, Chemiker und Techniker, die mit radioaktiven Stoffen umgehen, insbesondere das wissenschaftlich-technische Personal radiologischer Labors in den Kliniken, finden in dem Band sehr viele nützliche Informationen und sollten auf jeden Fall zu seinem Leserkreis gehören.

G. Brunner

Energy and Nuclear Power Planning in Developing Countries

Vienna: IAEA 1985. 814 S., 244 Abb., 155 Tab. (Technical Reports Series No. 245, STI/DOC/10/245); kart. ÖS 1600,—

Die Energiewirtschaft eines Landes gehört zu den permanenten Aktualitäten, die man planen kann und planen muß; denn nur so können energetische Bedrängnisse vermieden werden. Der Autor, Professor *Joan D. Stancesu*, legt in der Reihe der IAEA Technical Reports ein energiewirtschaftliches Lehrbuch vor, wie es bei interregionalen Trainingskursen über die Energieplanung in Entwicklungsländern unter besonderer Berücksichtigung der Kernenergie verwendet werden soll.

Das Lehrbuch ist in 4 große Teile gegliedert, die jeweils in insgesamt 18 Kapitel untergliedert sind. Energiebereitstellung und rationelle Energienutzung, ökonomische Aspekte der Energiewirtschaft, Stand und Tendenzen der Weltenergieentwicklung und Energieplanung sind die großen Teile. Dabei werden durchaus Nebenlinien in der Energiewirtschaft mit behandelt wie beispielsweise Fragen der Umweltbelastung durch die einzelnen Formen der Energiebereitstellung, Energietransport und Energiespeicherung oder auch Probleme der Vergasung und Verflüssigung von Biomassen bis hin zur Ethanol- und Biogasherstellung aus Biomassen, wie sie gerade für tropische Regionen von hohem Interesse sein dürften. Im Mittelpunkt stehen natürlich Fragen der Energieplanung und der Kosten-Nutzen-Analysen bei unterschiedlichem Einsatz der Primärenergieträger.

Das reichhaltig angebotene Zahlen- und Bildmaterial läßt erwarten, daß dieses Lehrbuch für Energiewirtschaftler und verantwortliche Planer im Energiesektor einen hohen Stellenwert haben wird. Hervorzuheben sind auch die insgesamt 945 Literaturstellen, die leicht eine weitere Vertiefung in dieses Arbeitsgebiet erlauben. Sie entstammen zum größten Teil den letzten Jahren (rund 60% aus den Jahren 1981—84), so daß auch in dieser Hinsicht das thematisch hoch befrachtete Lehrbuch seinen Nutzen anbietet. Es zeigt auch deutlich, in welch hohem Maße die IAEA Anteil daran hat, den weltweiten Energiebedarf durch den Erfahrungsaustausch auf internationalen Tagungen zu diskutieren. Der vorliegende Technical Report reiht sich zwanglos in dieses anhaltende Bemühen mit ein.

H. Hübner

AKADEMIE-VERLAG BERLIN

Physical Research

Physikalische Forschungsergebnisse aus aller Welt

Der Akademie-Verlag Berlin beginnt 1985 mit einem Pendant zu seiner Reihe Mathematical Research. Dem internationalen Trend, aktuelle Forschungsergebnisse in schnell erscheinenden Kleinstmonographien zu veröffentlichen, will die neue Serie Physical Research nachkommen. In ihren Bänden sollen Themen aus allen Gebieten veröffentlicht werden, die für die internationale physikalische Forschung relevant sind.
Die Titel erscheinen überwiegend in englischer Sprache und im Manuskriptdruck.

HARALD BÖTTGER / VALERIJ V. BRYKSIN

Hopping Conduction in Solids

1985. 398 Seiten — 107 Abbildungen — 1 Tabelle — DDR 88,— M; Ausland 88,— DM
Bestell-Nr. 763 411 5
Bestellwort:
Boettger / Bryksin 6863 engl.

Hoppingleitung als grundlegender Transportmechanismus in Festkörpern dominiert speziell in allen Materialien mit kleiner Ladungsträgerbeweglichkeit und wird seit Ende der fünfziger Jahre untersucht. Böttger und Bryksin unternehmen mit ihrer Monografie den ersten Versuch, den Hoppingmechanismus unter einem einheitlichen Gesichtspunkt zu beschreiben. Ausgehend von ersten Prinzipien wird die Theorie sorgfältig entwickelt, so daß das Buch auch als Einführung in die Theorie der Transportphänomene in ungeordneten Festkörpern gelesen werden kann. Da die wichtigsten Resultate aber ebenso mit Hilfe intuitiver Betrachtungen gewonnen werden, ist es gleichfalls nützlich für den an mathematischen Details weniger interessierten Experimentator.

HEINZ-EBERHARDT ALBRECHT

Laser-Doppler-Strömungsmessung

1986. Ca. 240 Seiten — 136 Abbildungen — 14 Tabellen — DDR etwa 45,— M; Ausland etwa 45,— DM
Bestell-Nr. 763 533 7
Bestellwort: Albrecht, Laser 6917

Diese Meßart wird bei der Strömungsanalyse in Forschung und Entwicklung zunehmend als modernes berührungsloses und quasipunktförmiges Verfahren genutzt. Albrecht vereint in seiner Monografie die Darstellung der physikalischen Grundlagen dieser Messung mit Erläuterungen zur Theorie des Meßsystems und zum Aufbau der Meßanordnung sowie zur Dimensionierung der Meßstrecke, so daß der Leser befähigt wird, einfache Meßsysteme selbst aufzubauen bzw. vorhandene besser zu beurteilen und optimal einzusetzen.

HELMUT HESS

High Pressure Plasmas

1986. 250 Seiten — Abbildungen — DDR etwa 40,— M; Ausland etwa 40,— DM
Bestell-Nr. 763 543 3
Bestellwort: Hess 6925

Die Untersuchung von Hochdruckplasmen hat einerseits Bedeutung für die theoretische Grundlagenforschung und ist andererseits zunehmend von technischem Interesse (Anwendung von Hochdruckplasmen in Speziallichtquellen von Schaltgeräten, Laserfusion, ...). In seiner Überblickdarstellung der Physik des Hochdruckplasmas (speziell des Niedertemperaturplasmas) wird Heß beiden Aspekten gerecht: Er verknüpft sorgfältige Aufbereitung der vorliegenden experimentellen Ergebnisse (auch eigener Arbeiten) und davon ausgehende kritische Wertung des erreichten Standes der theoretischen Beschreibung.

KLAUS BAUMGÄRTEL / KONRAD SAUER

Topics on Nonlinear Wave Plasma Interaction

1986. 250 Seiten — Abbildungen — DDR etwa 40,— M; Ausland etwa 40,— DM
Bestell-Nr. 763 544 1
Bestellwort: Baumgaertel / Sauer 6926

Plasma-physikalische Probleme werden im Titel Baumgärtel/Sauer, Topics on Nonlinear Wave Plasma Interaction dargestellt. Seit Anfang der 70er Jahre sind bei der experimentellen und theoretischen Untersuchung der Wellenausbreitung in Plasmen verstärkt nichtlineare Prozesse erforscht worden.
Stimuliert wurden diese Forschungen durch die Entwicklung leistungsstarker Laser und deren Einsatz zur Realisierung der Kernfusion. Baumgärtel und Sauer geben nun eine zusammenfassende übersichtliche Darstellung der wesentlichen nichtlinearen Erscheinungen auf der Basis der hydrodynamischen Plasmabeschreibung. An die theoretische Diskussion schließt sich jeweils eine Auswertung unter Einsatz numerischer Methoden an.

ISOTOPENPRAXIS

Band 22 1986 Heft 1

Contents / Inhalt

►Language of Publication ►Publikationssprache

Isotopenpraxis is indexed in Current Contents/PC & ES, Science Citation Index, and INIS Atomindex

FSC
www.fsc.org
MIX
Papier | Fördert gute Waldnutzung
FSC® C083411

Kolibri 360 GmbH
Ferdinand-Jühlke-Straße 7
99095 Erfurt, Deutschland
produktsicherheit@kolibri360.de